CASES IN CONSTRUCTION MANAGEMENT

A Construction News Book

By W. J. Slater, FCIOB

LONDON AND NEW YORK

Published 1988

Transferred to Digital Printing 2005

© Routledge

ISBN 1 85032 032 2

A Construction News Book

Routledge, 2 Park Square, Milton Park, Abingdon, Oxon, OX14 4RN

Contents

Case 1: Tyne Construction Ltd Case	7
Case 2: Jones Bros Ltd Case	14
Case 3: Environmental Services Ltd Case	20
Case 4: GM Construction Co Ltd Case	29
Case 5: The Scott Case	32
Case 6: The National Plant Co Ltd Case	42
Case 7: Northern Construction Ltd Case	53
Case 8: North East Development Ltd Case	61
Case 9: Discipline Cases	63
Case 10: The 4½ Day Week Case	65
Case 11: Proposed Consortium Case	66
Case 12: S Sands (North East) Ltd Case	73
Case 13: Lawnswood Housing Ltd Case	82
Case 14: UK Construction Ltd Case	84
Case 15: Dispute Case.	86
Case 16: Northumbria Construction Ltd Case	88

CONSTRUCTION MANAGEMENT CASE WORK

The use of cases in management education is now well established. There is, however, a gap in the number and quality of construction management cases. It is hoped this book will help to fill the gap.

The cases will be suitable for all management courses, both in-company and college based, but will be particularly useful for the Chartered Institute of Building Membership I and II examinations.

The cases have been tried out on students and found to be effective.

The author has been a lecturer and management consultant for 24 years. He has his own consultancy business and is also a Director in PES Associates – Management Consultants. The case work is principally taken from his own case book.

It is important to have a tutors manual as all the information need not and possibly should not be given to the student. Additional information which is likely to be called for is in the tutors manual. There is, however, still scope for the tutor to use his own imagination.

People learn better when doing and the case method is one of the best ways to do this. The tutor becomes a course manager and not a talking machine. He will learn as well as the students.

Geographical names may be changed to suit the location the case is being operated. Local conditions should be considered.

INTRODUCTION

The case study method of learning has been well tried. It has the following advantages:
1 Students need to think more about the subject;
2 The subject is brought to life;
3 Students realise that everything is not 'cut and dried';
4 It helps students to see the inter relationship of one topic with another;
5 They develop a deeper understanding of the subject;
6 They add interest to the learning situation which improves the learning process;
7 They develop an analytical approach to the subject;
8 The experience of the student is tapped.

Not all information is given in the case. Additional information should only be given if requested. Students must know what information they require to give a satisfactory answer.

The tutors text deals with the additional information students are likely to ask but there is still scope for the imagination.

The cases could be presented in verbal or written form. The larger the case the more likely the written form will be the best. Paper, felt pens, overhead projectors, etc should be available for presentation.

Students should be encouraged to present the case as this helps to develop their communication skills and gives them more confidence in standing in front of a group. Members of the group should share the presentation.

The time taken for each case will vary with the type of student. A degree of judgement will need to be used but don't relinquish quality for quantity.

Questions should be encouraged and revision may need to be given to individuals if they do not understand any aspect of the information given.

Groups should not exceed five as some people will not have the opportunity to contribute.

Groups should be encouraged to constructively criticise the presentation of other groups and be prepared to ask and answer questions.

It must be emphasised that in a case study there is not one answer but an answer must be found. Most situations in practice are of this nature.

Sometimes information given may be irrelevant but students will be expected to sort this out.

Tutors should avoid ready made answers. The cases used in this book are based on real life situations but they have been adapted to enable them to be carried out in a reasonable time without losing any of their value. Names used are fictitious to avoid companies being identified. the author does not wish to create any embarrassment to any organisation or individual.

Case 1
TYNE CONSTRUCTION LTD

You have been approached by a firm of 'head hunters' to take over the position of Managing Director of Tyne Construction Ltd. The salary offered is generous with the usual fringe benefits.

After looking at the company you have decided to accept the position provided the company give you a three year contract. If at the end of the three year term you have proved satisfactory a further five year contract will be given.

The history of the company is that it was started as a small private house building firm in 1933 by a Mr George Anderson. By the early sixties the company employed 1,000 people and was engaged in Civil Engineering as well as building and small works.

The firm was incorporated as a private limited company in 1947. Mr George Anderson died in 1970 and the company was inherited by his wife and four children (two daughters and two sons). She did not re-marry. The share-holding is as follows:

     Mrs Anderson 60%
     Each child 10% each

The children are not interested in the business and do not want to get involved apart from collecting dividends. Mrs Anderson is a non-executive director to the company and draws a remuneration of £6,000 per annum. She is 75.

Mr George Anderson had appointed a first class manager and the business was quite successful until he died five years ago. A new manager was appointed but has recently been dismissed.

The profit/loss before tax for the last six years is as under:

Year 1 (last year)	(£150,000) loss
Year 2	£200,000
Year 3	£150,000
Year 4	£500,000
Year 5	£800,000
Year 6	£1,000,000

Turnover for the last six years is as under:

Year 1 (last year)	£25,000,000
Year 2	£30,000,000
Year 3	£26,000,000
Year 4	£24,000,000
Year 5	£22,000,000
Year 6	£20,000,000

The turnover is calculated by taking value of certificates and invoices plus closing work-in progress minus opening work-in-progress.

Work-in-progress is valued at cost.

The balance sheets for the last five years is shown in Appendix I.

The bank overdraft facility five years ago was £500,000. It was increased to £2 million four years ago, £4 million three years ago, £6 million two years ago with £1 million colateral put up by the family as a personal guarantee. It was increased to £7 million last year. The bank are putting a great deal of pressure on to get it down to £4 million.

This has caused Mrs Anderson a great deal of concern and after a meeting with the family decided to replace the Managing Director as a first step towards recovery.

The present organisation is shown in Appendix 2.

The breakdown of turnover last year was:

Civil Engineering	£6,000,000
General Building	£15,000,000
Small Works	£4,000,000

Breakdown of profit/loss:

Civil Engineering	£100,000
General Building	(£650,000) Loss
Small Works	£400,000

The work the company has on for the current year is:

Civil Engineering	£4,000,000
General Building	£8,000,000
Small Works	£2,000,000

Any building jobs under £200,000 are classified as small works. Jobs could be as low as £200 in this section.

The geographical area is Northern counties, including Yorkshire and Lancashire. The company does not operate in Scotland.

The company has a small branch office in Leeds and one in Keswick.

TYNE CONSTRUCTION LTD CASE 9

Each of these offices is headed by one of the contracts managers. Each one has a small staff of four.

The People
Financial Director
Age 45. Married with two children. Is a qualified management accountant. He has been with the company eight years. Did not get along with the last Managing Director. He is keen to stay with the organisation. Salary £14,000 plus Company car and fringe benefits.

Surveying Director
Joined the company five years ago. Was a friend of the last Managing Directgor. Qualified IQS. Age 58. Married – no children. Salary £14,500 plus car etc.

Construction Director
Ex-joiner, site manager, contracts manager. Age 52. Married – one child. Salary £16,000 per annum plus benefits. No qualifications. Has not been on any short courses. He has been with the company six years. Would like to stay with the company. Hardworker.

Plant Manager
He has been with the company three years. Previously worked as assistant plant manager with large national company. Age 48. Married – no children. Salary £13,000 per annum. Reasonable worker.

Small Works Manager
He has worked all his working life with the company. 48 years old. Married with two children. Salary £12,000 per annum. Ex-joiner by trade. Very enthusiastic. Good knowledge of the construction process. Does not think management is as good as it was but will not comment further.

The following points have been noted:
1. Company is not claims conscious.
2. Contract planning is very simple and generally not effective.
3. The bonus scheme collapsed three years ago.
4. Wastage on site is high.
5. Site managers think there could be a big improvement in buying, plant (quality very poor) and planning.
6. The company had a good reputation for quality but this has taken a knock with two bad examples of bad workmanship.
7. Cost control is primitive.
8. Tender success rate last year was 1 in 3.
9. Industrial relations is very good.

10. The Company is a member of the Building Employees Confederation and Federation of Civil Engineering Contractors.
11. Most of the men are in a union.
12. The safety record is poor. Two fatal accidents in four years and 10 accidents resulting in absence of more than three days last year.
13. Plant is in poor condition.
14. The company owns 15 cars. The value is included in the plant figure shown in the balance sheet.
15. The loan is due for re-payment in five years time.
16. The debtors are all good.
17. The stock is valued on the FILO system.
18. 30% of the work is sub-contracted in the civils section;
    40% of the work is sub-contracted in the building section;
    15% of the work is sub-contracted in the small works section.
19. The company would consider going public if this was possible. This was first discussed five years ago but no action taken.
20. The company employs its own labourers, joiners, bricklayers, plasterers, machine operators, plumbers and concretors.
21. The buying procedures are poor.
22. The company has a small joiner shop which produces mainly for the small works department.
23. Estimating procedures need revising.

As Managing director you will have a free hand on running the company but the family would like you to produce a report on how you intend to get out of the present difficulties. Your report should cover both short and long term aspects.

The type of work the company has carried out is as follows:

Hospitals	Police Stations
Schools	Swimming Pools/Leisure Centres
Fire Stations	Supermarkets
Welfare Centres	Dock Work
Re-furbishment	Commercial Buildings
Re-vitalisation for Local Authority	Shopping Centre
Local Authority Housing	Railway Civils Work
Sewers	Farm Buildings
Industrial Work	Sheltered Accommodation
Small Road Contracts	Extension to Buildings

The company has not built private houses since the last war.

Turnover of office staff is high as the company does not have a good reputation for salaries.

The head office accommodation could take a 50 per cent increase.

The company does not own computers.

There is a small canteen in the head office. Meals are subsidised by the company to a maximum of 30 per cent.

There is a staff association but it is poorly attended and is likely to cease functioning altogether.

The Personnel Officer has one Training Officer working for him. The training officer has little knowledge of construction and it is generally thought he got the job because of his friendship with the previous Managing Director. He has been in post for two years.

The only training the company appears to do is with apprentices. The present value of the land and buildings which consists of the head office, joiners shop and yard is £350,000.

The company does not usually find out if it has made a profit or loss on a contract until after the job is finished.

The Quantity Surveying Director does not think feed back methods are cost effective.

The company is finding it increasingly difficult to obtain work.

The company policy is not to use labour only sub-contractors.

Appendix I

Note: Year 5 is last year

As at 31 May	Year 1 £	Year 1 £	Year 2 £	Year 2 £	Year 3 £	Year 3 £	Year 4 £	Year 4 £	Year 5 (last year) £	Year 5 (last year) £
Fixed Assets										
Land/buildings		200,000		200,000		200,000		200,000		200,000
Plant less depreciation*	800,000		600,000		450,000		500,000		380,000	
Office furniture	300,000	1,300,000	225,000	1,025,000	168,750	818,750	126,560	826,560	94,920	674,920
Current Assets										
Work in progress	6,000,000		7,700,000		9,600,000		10,000,000		12,500,000	
Debtors	2,200,000		3,600,000		4,800,000		6,000,000		5,500,000	
Stock	400,000		800,000		900,000		800,000		700,000	
Cash in bank	100,000	8,700,000	—	12,100,000	—	15,300,000	—	16,800,000	—	18,700,000
		10,000,000								
Current Liabilities										
Current tax	300,000		200,000		120,000		40,000		50,000	
Creditors	390,000		5,000,000		6,300,000		6,000,000		6,500,000	
Bank overdraft	—		1,625,000		3,548,750		5,386,560		6,824,920	
Proposed dividend	100,000	4,300,000	380,000	7,205,000	110,000	10,078,750	150,000	11,576,560	NIL	13,374,920
Net assets		5,700,000		6,120,000		6,040,000		6,050,000		6,000,000
Financed by:										
Ord share cap (£)	1,000,000		1,000,000		1,000,000		1,000,000			
Revenue reserve	1,500,000		2,000,000		2,000,000		2,000,000		2,000,000	
Deferred Liabilities										
Loans	3,000,000		3,000,000		3,000,000		3,000,000		3,000,000	
Future tax	200,000	5,700,000	120,000	6,120,000	40,000	6,040,000	50,000	6,050,000	NIL	6,000,000

*includes small tools

Appendix II

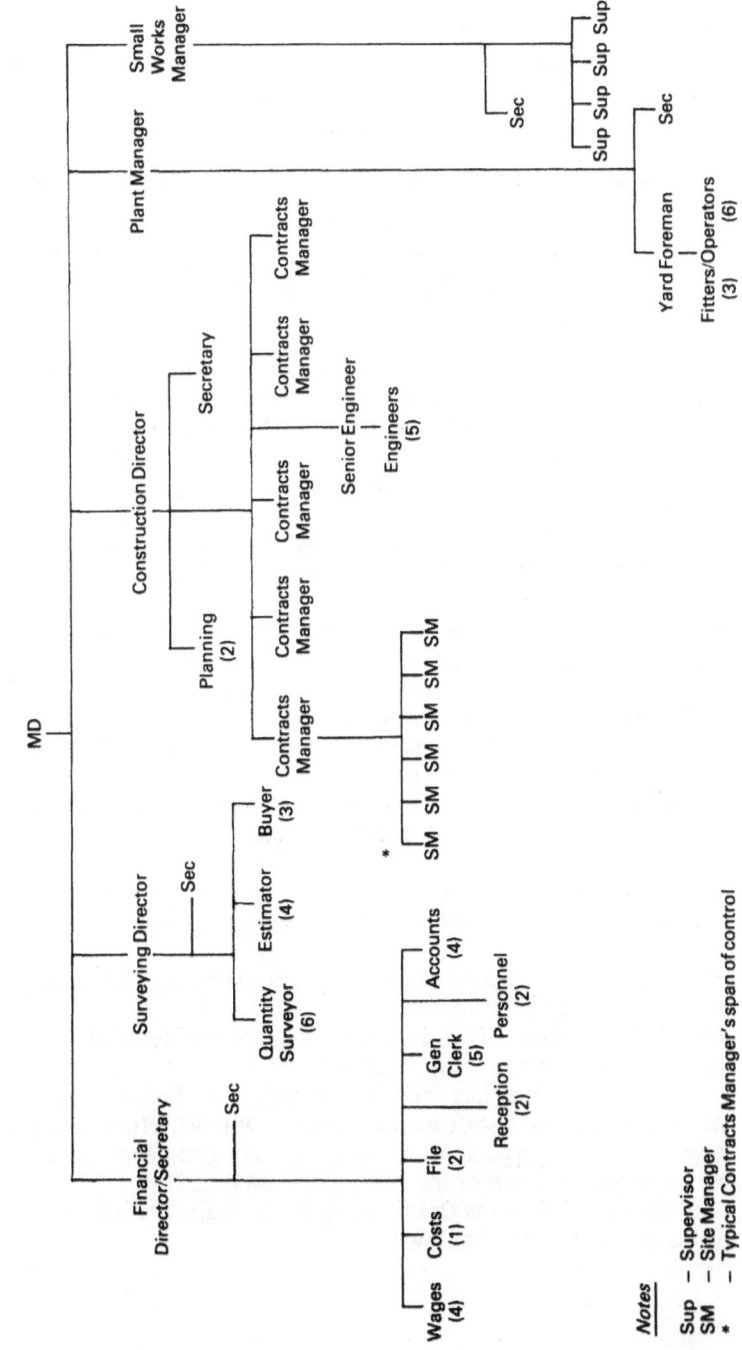

Case 2
JONES BROS LTD

Jones Bros Ltd are medium sized builders operating in the North East. They employ, on average, 600 men. The policy of the company is to restrict its tendering policy to jobs under £2.5 million. It has a jobbing section and a small works department. Its turnover is approximately £18 million. It has at present 38 contracts, ranging in size from £150,000 to £2.2 million plus jobbing, plus 14 small contracts under £100,000.

The company does not specialise in any particular type of work but sets its geographical limitations south of the Scottish border and north of Yorkshire.

It employs an indirect staff of 100 including head office and site staff.

John Jones has asked you to advise him on the organisation of the company. He would like to expand to 1,000 men but feels that all is not right in the business. When asked about this he summarized his feelings as:
1. The profit on turnover hasn't exceeded 2.5 per cent over the last five years and last year was 0.5 per cent.
2. The total capital employed is nearly £4 million and Jones does not feel that it is being worked hard enough. Turnover last year was £18 million.
3. Cost control in the company is good but financial control is weak. Jones feels that he would like a lot more information to base his management decisions on.
4. There appears to be a lot of crises and problems on site which are costing the company a lot of money.
5. Communications are bad. Policies which are decided at the top are not getting down to the site in an effective way.
6. Office staff turnover has been unduly high in the last two years. He has lost some very good staff such as Jack McDonald, a first class planning engineer and George Scott a young but very good accountant.

An organisation chart of the company is attached.

You have discussions with members of the staff. Details of these discussions are summarized below:

JONES BROS LTD CASE

George Baker *(Contracts Director)*
Personal friend of Jones. Brought up together as children. Both families meet a lot socially. Both Baker and Jones entered the business about the same time. Very forceful man, rather domineering. No education since leaving Grammar School at age of 16 with School Certificate. No technical education. Hasn't attended any course during the last 12 years. Great capacity for work. Commences work at 8.00 am. Never leaves office before 6.30 pm. Often works until 9.00 pm.

Q How often does he visit each contract?
A Depends on type of job but on average once per fortnight.
Q Does he feel that he has the job under control?
A Yes.
Q Does he feel overworked?
A Yes, but I like hard work.
Q What does he attribute to the low profits on contracts?
A Bad foremen – you cannot get good foremen these days. Men don't want responsibility.
Q Why did Jack McDonald leve?
A He got more money at a Newcastle builders.
Q What is his impression of the Office Manager?
A Ian Todd is a good company man. He has been with the company almost since it commenced. He is moody. He works hard enough.
Q What kind of planning techniques does the company use?
A Long term bar charts. They are good enough in general, but rarely work out in practice.
Q Does the company use network planning?
A No – I think it is a lot of nonsense.
Q Do you know anything about network planning?
A A little.
Q Have you ever attended a course on the subject?
A No.
Q How do you know anything about it?
A I read it in books.
Q Could McDonald do network planning?
A He could – in fact I think he was a bit of an expert. I told him we didn't use that nonsense here.
Q Have any of the foremen attended a course on network planning?
A Two of them went to a course run by a bloke called Slater.
Q Have they had a chance to practice it?
A No.
Q How does he spend most of his time?
A Sorting out problems – a general trouble shooter.

Harry Taylor *(General Foreman)*
Harry has worked for Jones for 12 years.
Q What does he think of the firm?
A It isn't a bad outfit to work for, but there are problems. I like John Jones – he is a nice fellow, but George Baker can be a bit of a swine at times. It doesn't bother me much because we hardly see him unless the job is losing.
Q How often do you see him?
A On Average, once every three weeks.
Q Do you get much support from head office?
A Practically none – we have to sort out most of our problems ourselves – this can be very difficult at times as it is difficult to get to the right person.
Q What kind of problems occur the most?
A Shortage of material, wrong material delivered, non-arrival of sub-contractors. Changes in drawings and clients instructions – often received after that part of the job is done. Labour turnover and bonus problems.
Q Would he like more support from head office?
A Support – yes. Interference – no.
Q Why do the contracts have such a low profit return?
A Bad management at the top.
Q Have you been on any courses recently?
A I went on a three-week site management course. It was a good course but I would like to get more opportunity to put the things taught on the course into practice.
A What kind of things?
A Work study, planning and plant selection. The company already has a good costing system. They sent the wrong man to the course – it is top management who should have been there.

Ian Todd *(Office Manager)*
Duties include supervising office staff, producing financial reports when requested and general office routine.
Q You have a good costing system – who is responsible for it?
A I generally supervise it, but George Potts installed it and operates it. He is a wizard at the job. He is under me. He is 28 years old and has been with the firm four years. He has a DMS and is a Chartered Accountant.
Q What experience have you had?
A I joined the firm when I left school at 14, that was in 1944. I was employed by John's father as an office boy and eventually became office manager.
Q Have you attended any course at a college or the like?

JONES BROS LTD CASE 17

A No, but I have read a lot.
Q Do you feel you are up to date in your accounting, particularly the management accounting field?
A I am not bad – I suppose there is always room for improvement.
Q Do you use budgetary control?
A No.
Q Do you produce a balance sheet every month?
A You must be joking – it takes us all our time to get one out each year.
Q Do you operate financial ratios or do you have any method of controlling turnover?
A No.
Q Do you use marginal costing?
A Tell me what marginal costing is and I will tell you if we use it.
Q Have you ever joined an inter-firm comparison scheme?
A We had an advert around but I thought £200 was a bit too much to tell us what we know about ourselves. There appeared to be too much work in it anyway.
Q What financial techniques do you use?
A Traditional book keeping and monthly cash flows. I am working on an idea to speed up jobbing accounts.
Q How do you get on with your staff?
A Not too bad, I don't take a great deal of notice of them unless they are not doing their job.
Q How much do you get paid?
A That is my business not yours (you check up and find he gets £14,600 per year).
Q Do you use much office machinery apart from typewriters?
A We have a duplicator and two mechanical comptometers.
Q What pay system do you use?
A We have a wages book or I should say books and copy the details onto a packet. Four girls do this full time.
Q How many bonus clerks do you employ?
A Three.

Mary Smith *(Office Clerk)*
Q What job do you do?
A Filing and I enter up the purchase journal.
Q Have you got too much work to do?
A You must be kidding. I get bored for half of the week.
Q Would you like to take on more work?
A I suppose I would.
Q How many girls feel like you about the job?
A I would say all of them.

John Jones

Q   Why do you not employ contracts managers?
A   Baker says he can manage, but I would like to see more contact between sites and head office. I stroll around some sites and I get hit with a mass of problems.
Q   Do you have regular meetings with the foremen?
A   We don't have any meetings with all of the foremen together.
Q   What are your feelings about Ian Todd the office manager?
A   I inherited Ian. He is a loyal, hardworker, very honest, very punctual and often miserable. I really can't make my mind up about Ian.

You tried to have a meeting with John Butterfield, the small works manager, but up to date this has been impossible to arrange. He says he is too busy. You have carried out an investigation into Butterfield and find him to be a competent small works supervisor. He has no potential to be in a higher position than he is now. He has a good knowledge of the construction process. His section is probably making money.

What would you advise?

JONES BROS LTD CASE

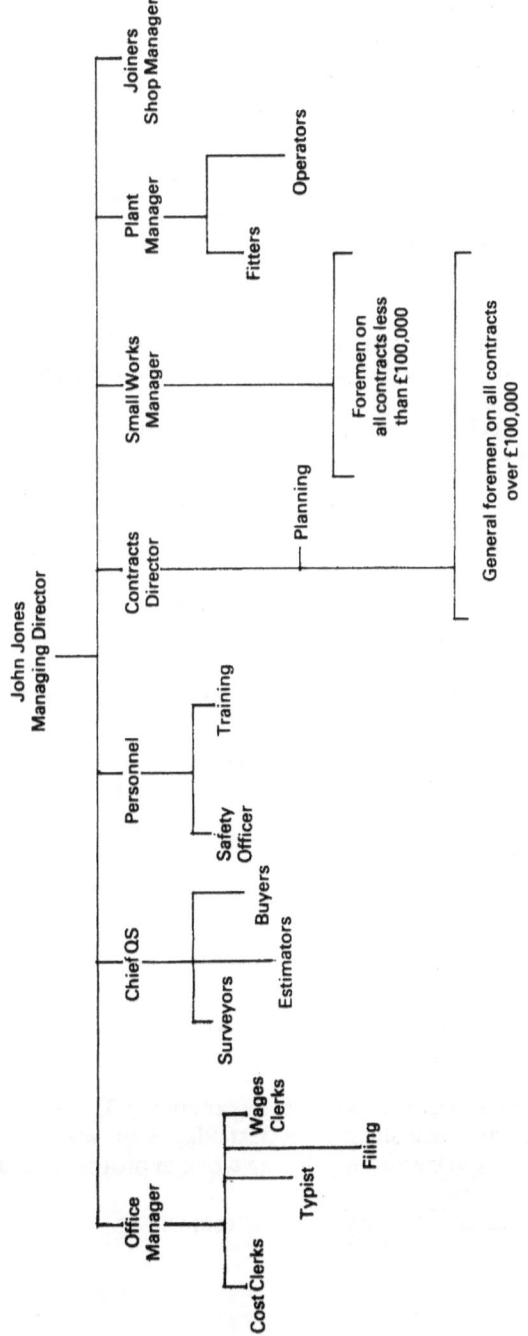

Case 3
ENVIRONMENTAL SERVICES LTD

You have been appointed managing director of Environmental Services Ltd with a good salary, fringe benefits and a profit bonus. The previous managing director J. Black has been dismissed.

Environmental Services Ltd is a subsidiary of a large multi-national construction organisation. The subsidiary specialises in all aspects of building services including design, sub-contracting and research for external organisations.

E.S. Ltd (referred to as the company) commenced trading 10 years ago and for the first five years made good progress under the management of J. Stone. Mr Stone died five years ago and his position was filled by a 'head hunting' organisation who appointed John Black who has been dismissed.

The total turnover of the company for the past 10 years is as under:

	(millions)
Year 1 (last year)	£14.2
Year 2	20.1
Year 3	25.2
Year 4	26.3
Year 5	24.4
Year 6	22.2
Year 7	20.1
Year 8	18.6
Year 9	10.7
Year 10	3.1

The Trading Profit and Loss account for year ending 31st May is shown in Appendix A. The balance sheet as at 31st May is shown in Appendix B. Turnover is calculated by adding closing work in progress to sales less opening work in progress.

Profit before tax for the last 10 years is as under:

	£
Year 1 (last Year)	14,000
Year 2	32,000
Year 3	65,000
Year 4	110,000
Year 5	200,000
Year 6	2,200,000
Year 7	2,000,000
Year 8	1,800,000
Year 9	400,000
Year 10	250,000

The present organisation chart is shown in Appendix C.

Accommodation
The company operates from an office block within five miles of Newcastle. The office block has four floors. The ground floor is occupied by reception, laboratory, mail room and accounts. The remaining floors have an occupancy of 75 per cent. There is a 2,000 m^2 car park. A single storey store area is located in the site with a floor area of 1,000 m^2. Operatives accommodation is adjacent to the stores block.

Public transport to the office site is very good. The company own a mini-computer which is only used for accounts and payroll. There are four micro-computers in partial use.

Geographical limitations
The company operates North of a line from the Humber to the Scottish border plus an area 30 miles radius of Edinburgh. The company has branch offices in Edinburgh, Leeds and Manchester. There is a regional manager with staff in each of the branch offices. All design, estimating, accounts, payroll, buying and planning is carried out at the head office.

Company Car Policy
Company cars are given to top line executive and contract supervisors. Pool cars are available for the remaining staff as and when required.

Research Department
The research department occupies a well equipped laboratory and is headed by John Windsor age 45. Windsor has a PhD and would be described as highly intelligent but non-commercial.

The department carries out work for the company but the bulk of its work is for other organisations. Much of its work is in dealing with unusual building services problems which occur or are likely to occur in building, off-shore oil platforms, process engineering works and ships.

The department was set up five years ago and since then has never made a profit. Losses each year have been small ranging from £20,000 to £40,000.

Windsor argues that his department is a loss leader and may bring in or brings in other work for the company. There isn't a lot of evidence of this taking place. The department appears to be kept busy but managing research appears to have different problems to managing other parts of the company.

The charge out rates for research are at present £23 per hour for research assistants and £15 per hour for a technician. This includes overheads and is supposed to carry an element of profit. Materials used are charged at cost plust 25 per cent.

Further investigation into the company's activities reveal:
1. The company is not 'Claims' conscious.
2. Planning is carried out by bar chart and is rarely up-dated.
3. Material wastage appears to be high.
4. Cost control is primitive.
5. Industrial relations are good.
6. Safety has a poor record. Last year one fatal accident, 10 accidents involving several weeks away from work.
7. Most of the operatives are in a union.
8. Debtors are all good.
9. Plant and vehicles are generally in poor condition.
10. Stock is valued on the FILO system.
11. The buying procedure is not effective.
12. Branch managers say they are not kept in the picture.
13. Estimating procedures need revising.
14. Communications are not very good.

The type of work the company has carried out is as follows:

Hospitals	Swimming pools/leisure centres
Schools	Supermarkets
Fire stations	Commercial buildings
Welfare centres	Shopping centres
Refurbishment	On-shore gas installations
Revitalisation	Railway station metro (2)
Industrial work	Sheltered accommodation
Police Stations	Civic Centres

Apart from research work and one on-shore gas installation the company has not been involved in process engineering work.

There is a small canteen in head office. Meals are subsidised by the company up to 30 per cent.

There is a staff association but is poorly attended and is likely to cease functioning altogether. The company subsidises this to the extent of £40,000 per year.

ENVIRONMENTAL SERVICES LTD CASE 23

The surveying director does not consider feed back methods are cost effective.
The company is finding it increasingly difficult to obtain work.
The company policy is not to use labour only sub-contractors.

Staff
The financial director, Eric Todd, commenced with the company five years ago. He had worked for the same company as the previous managing director who brought him with him from his previous company. He has no recognised qualification and commenced his working life at the age of 14 as an office boy. He says he has learnt the hard way. He is now 56 years old.

You have interviewed Todd and the following is an outline of the interview.
Q   Do you produce monthly balance sheets and profit and loss accounts?
A   You must be joking – it takes all our time to get out the annual accounts.
Q   Does the company use marginal costing?
A   No.
Q   Does the company use ratio analysis?
A   No.
Q   What does the man on statistics do?
A   He prepares statistics for Government returns and for internal use. He also helps in the preparation of information for the Registrar of companies. Actually he is my brother-in-law. He is a big help.
Q   What techniques does the company use to control turnover?
A   None.
Q   Are the debtors too great?
A   I think they are reasonable.
Q   Is work in progress too high?
A   I would say no.

John Elliott Age 46 *(Technical Manager)*
Q   What is your job?
A   I am not quite sure.
Q   What do you do?
A   Trouble shoot technical problems. Give advice if necessary to design engineers. Co-ordinate services.
Q   Have you any major problems?
A   My area is too big. I haven't the time to do anything well. I get very frustrated.
Q   How would you rate the company from a technical aspect?
A   Good but not as good as they used to be.
Q   Why is this?

A The staff are not as good and the leadership of the company left a lot to be desired.
Q Do we have any technical problems?
A Yes lots, we have a big one at the moment. We designed the services for a large Civic Centre. The building is too hot in the summer and too cold in the winter. The air conditioning is playing up quite a bit, we have had one flood which put the electrical services out of action for three days because a fail-safe switch did not work. There could be a large claim against us.
Q How much?
A Approximately £180,000.
Q What was the main cause?
A Not enough information – Not enough time – Not enough staff.
Q What would you do to improve the company?
A I would need six weeks notice to answer that one.
Q What qualification do you have?
A Degree in Building Services Engineering from Newcastle Polytechnic.

Jack Butterfield Age 50 *(Buyer)*
Q Do you get the best prices available?
A Yes.
Q How do you know?
A Experience.
Q Do you spend much time interviewing reps?
A I avoid them like the plague. They are a dam nuisance.
Q Do you work in conjunction with the estimating dept?
A I generally get some quotes for them.
Q What is the documentation like?
A The usual stuff – requisition, orders, quotes etc.
Q Were they designed for this company?
A No.
Q Do you make use of the computer?
A No – I haven't the time.
Q When did you last look at the whole of the buying procedure?
A I haven't. They just grew as the need arose.
Q Are you short staffed?
A Yes. We need to spend a lot of time chasing up suppliers.
Q Why is this?
A Not sure, it would help if the orders were placed more in advance. We appear to work on the crisis method.
Q Do you work well with the planning section?
A I don't work with them at all.
Q Why not?

ENVIRONMENTAL SERVICES LTD CASE 25

A   I haven't found any need. The programmes produced aren't much good anyway.
Q   How do you know?
A   It is just what I hear?

Frank Smith Age 54 *(Chief Estimator)*
Q   What is our tender success rate?
A   Last year – 1 in 3.
Q   Do you have any feed back on the estimating?
A   No.
Q   Could there be more co-operation and co-ordination between the various departments?
A   There couldn't be any less than there is at present. Anything will be an improvement.
Q   Have you enough staff?
A   No.
Q   Do you use the computer?
A   No.
Q   What about the micros?
A   I haven't found any use for them.
Q   Who introduced the present estimating system?
A   No-one, it just grew.
Q   How long have you worked for the company?
A   Five years.

F. Jones Age 50 *(Construction Director)*
Q   How many supervisors are accountable to you?
A   All of them.
Q   How many is that?
A   Varies – 11-15.
Q   Is this span of control too big?
A   No, I can manage it. I like hard work.
Q   How often do you get onto the jobs?
A   I visit each job at least once.
Q   What is the planning section like?
A   Good enough.
Q   What techniques do you use?
A   Bar charts – but they don't often work out.
Q   Why?
A   Too many variations. Data is incorrect. Jobs keep changing.
Q   Do you think we should do more planning?
A   I don't think it will help.
Q   Does the planning work in conjunction with the surveying department.

A   No. I don't see where they can help.
Q   What is the standard of the supervision?
A   Some are good. The rest mediocre. Good supervisors are difficult to come by.
Q   Do we do any training with them?
A   No. I don't think it will help. they have either got it or they haven't.
Q   What is the turnover of supervisors like?
A   Quite high. I have changed five this last year. They may leave, or I may sack them. A good clean out now and again isn't a bad thing.
Q   Do you know what the cost of this turnover is?
A   No. I don't suppose it is very much.
Q   How long have you been with the company?
A   Ten years. I commenced life as a heating engineer and worked my way up in other companies before I came here.
Q   What qualifications have you?
A   City & Guilds Final.
Q   Do you think the company could be improved?
A   I don't think it is too bad now.
Q   What about low profits?
A   That is because of the recession. Times are hard now. I do think we should try to get more work. It is slackening off quite a bit.
Q   What does the company do to get new work?
A   It relies on its reputation, experience, contacts and adverts.

Your brief is to improve the company, turn the trend of poor profits. Re-organise when you feel necessary. Produce a plan of action for the short term (one year) and the long term (five years). The report should cover all aspects of the business.

ENVIRONMENTAL SERVICES LTD CASE

Appendix A

Trading Account for year ending 31st May

	£		£
Wages	4,586,000	Sales	10,200,000
Materials	6,200,000	Closing work in progress	8,000,000
Plant Hire	500,000	Closing stock	200,000
Opening work in progress	4,000,000		
Opening stock	100,000		
Gross profit C/D	3,014,000		
	£18,400,000		£18,400,000

Profit and Loss Account for year ending 31st May

	£		£
Directors remuneration and salaries	1,900,000	Gross Profit B/D	3,014,000
Interest	150,000		
Head office overheads	950,000		
Net Profit C/D	14,000		
	£3,014,000		£3,014,000

Appropriation Account

Tax	4,200		14,000
Reserves	9,800		
	£14,000		£14,000

Appendix B

Balance as at 31st May

Authorised Capital 2 m	£1 ord share		
	£		£
1,000,000 £1 ord shares	1,000,000	Office Building/Land	1,000,000
Reserves	3,000,000	Plant and vehicles less	
Loan from Parent Co	6,000,000	depreciation	400,000
Bank overdraft	1,500,000	Debtors	6,000,000
Future tax	4,200	Work in progress	8,000,000
Current tax	12,000	Stock	850,000
Creditors	4,733,800		
	£16,250,000		£16,250,000

ENVIRONMENTAL SERVICES LTD CASE

Appendix C

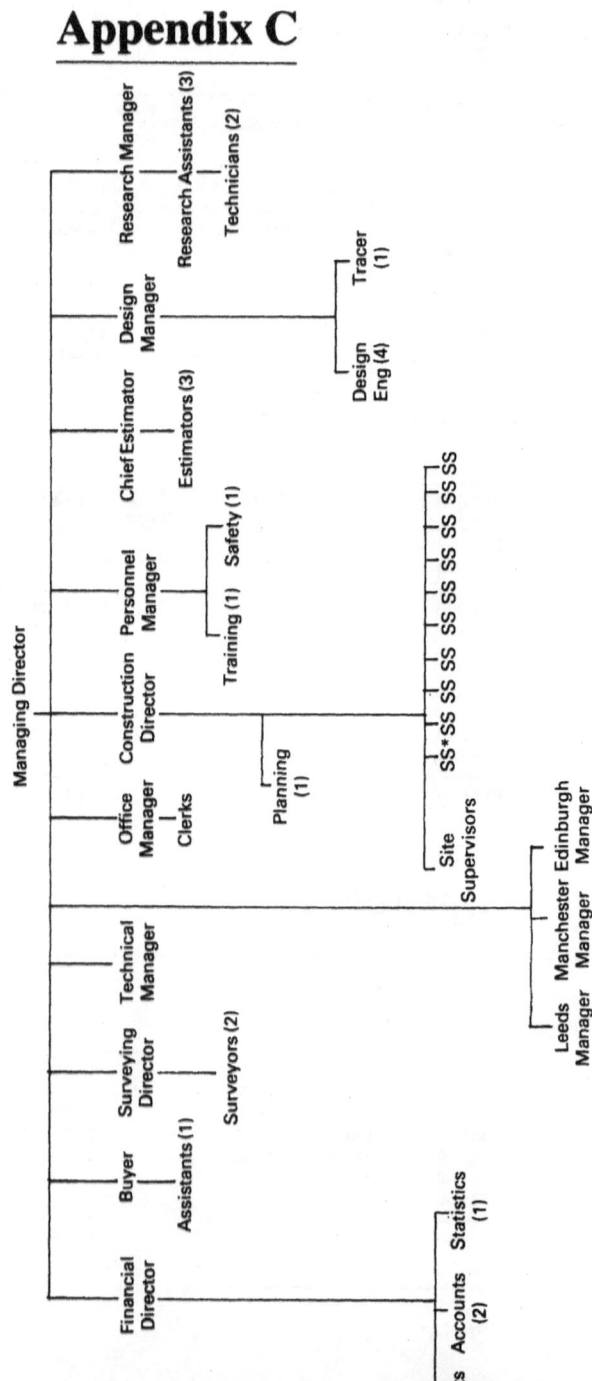

*Site Supervisors

Note: Number in brackets indicate number of staff.

Case 4
GM CONSTRUCTION CO LTD

You have been employed as a consultant to the GM Construction Co Ltd to install a material control system. You have visited the Company and discussed the problem with various key personnel.

The Company
The Company employs 200 men directly and approximately 30 per cent of its turnover is done by sub-contractors. Approximately 60 per cent of its turnover is in building contracts, 10 per cent on jobbing and small works and 30 per cent on private speculation housing.

The company operates mainly in the North East and has a turnover of approximately £9 million. at present it has 24 contracts or speculative house developments on ranging from £2 million to £100,000 plus jobbing work.

Discussion points are as follows:

Jim Broadway *(General Manager)*
1. Material costs are generally higher than estimated.
2. There appears to be a general lack of control as far as material is concerned.
3. He is not sure whether this is due to excessive waste, bad buying, bad estimating, pilfering (internal) pilfering (external) or being overcharged on invoices or a combination of all.

George Smith *(Buyer)* He:
1. Has one young assistant.
2. Takes off all the requirements from drawings and sends copy of order to site when the goods arrive (to enable site manager to check and to inform him that they have arrived).
3. Meets representatives from merchants twice per week for two hours. They take up too much valuable time says Smith.
4. He blames sites for too much waste and pilfering.

5. He also blames estimating department for not allowing enough for materials in the estimate.
6. He complains he has too much work to do.

Jack Brown *(Estimator)*
1. He complains that he doesn't get enough information from the buying department when preparing an estimate.
2. He too blames the sites for poor material control.
3. When asked how much cement should be used on the Wallsend School contract he says this would take him about five hours to find this out as the concrete materials are priced as a lump sum.
4. He hadn't heard of method statements.
5. On asked about tender success rate he replied that they picked up about one in five jobs. They did mainly schools, clubs, pubs and LA housing. They managed to get a few industrial buildings, but not many. They made a profit on some jobs and lost on others. The average profit generally worked out at 2 per cent.
6. He has no feed back system fromsite except to find out if the job as a whole made a profit or a loss.

George Green *(Site agent)*
George Green is one of Durham's senior site agents:
1. He says that the estimator must not allow enough for materials in the first place.
2. He complains that many items that the buyer takes off are incorrect and have to be returned increasing the cost.
3. He says he is unable to check until the materials actually arrive as that is when he receives a copy of the order. He reckons the site should take off most of the materials but he hasn't the staff to do it. That is what I used to do at Ainkeys Ltd and it worked better there.
4. How does he keep a check on waste? I keep my eye open for careless use of material and also instruct my foreman to do so, he says.
5. Asked how he knew if he was using too much cement in the concrete he replied that it should be alright as he always used a silo. The silo had not been checked in during the last six months.
6. He blames the men for pilfering, especially hardware. He couldn't see how this could be avoided.
7. Asked if he spot checked bricks, ready mixed concrete etc his reply was 'no'.
8. Did he have a site clerk? – No.
9. Value of the contract £1,995,000.
10. Did he employ a checker? – No, they cost too much money for the work they do.
11. Is his site typical of CM Ltd? – Yes.

GM CONSTRUCTION CO LTD CASE

12. Who signs delivery tickets? – Anyone who happens to be around.
13. Do all orders from his trades foreman go through him? – No, I believe in delegation, they are sent direct to Head Office.
14. Did he make regular inspection of plant such as crane skips – not regular, but I keep my eye on them.
15. What security did they have? – A watchman – 70 years old. He has a relief at the weekend.
16. How many site 'Break ins' had there been – six in 12 weeks.
17. Did they lose much with the 'break ins'? – Not much, to the best of my knowledge.
18. Did they have a time keeper? – No.
19. Did they check lorries leaving the site? – No.
20. Who supplied the sub-contractor's scaffold? – They supplied their own.
21. Did he give the QS much assistance in the monthly valuation? – No, that's his job.
22. He complained about serious delivery delays. What did he do if a material was delayed – he complained to the buyer.
23. Did he employ a 'picker-up' – No but periodically he put a gang of labourers on if they were slack.

George Blake *(Accountant)*
1. Did he or someone else check the invoices? – Yes as far as arithmetic is concerned.
2. Did they check with delivery tickets? – generally, but not always as they often didn't arrive off the sites.
3. Did the sites use a weekly goods received book or the like? – No.
4. Did he have a system of checking if crates returned were credited and did he check if crates were returned? – He generally checked this, but had no real system.
5. Did the Company pay its accounts promptly? – No, we always take an extra months credit.
6. Did you always lose the cash discount? – Not always but generally.

   Whaty would you do. Your report should indicate the points you would look for on site where material could possibly be wasted.

Case 5
THE SCOTT CASE

In a company which specialises in the construction of flats, houses and general building work, Ronald Scott has recently been promoted to the position of General Construction Manager. He is in charge of seven contracts in the Birmingham area and all of the Regional Managers.

The Construction Division is part of a National Company which also has interest in plant hire, builders' merchants, portable site accommodation, pressure jetting and joinery.

Scott was promoted three months ago by Norman Scott who is the Managing director of the main Board. Scott's predecessor has moved to another company after an approach by a head-hunting organisation.

Scott is a methodical worker and slow at making decisions. He frustrates his sub-ordinates because of his indecisiveness. He has an introvert personality. He is a good committee man but lacks self confidence when on his own. He likes to take time to ponder over things. He is intelligent, not forceful, lacks drive and is not very imaginative.

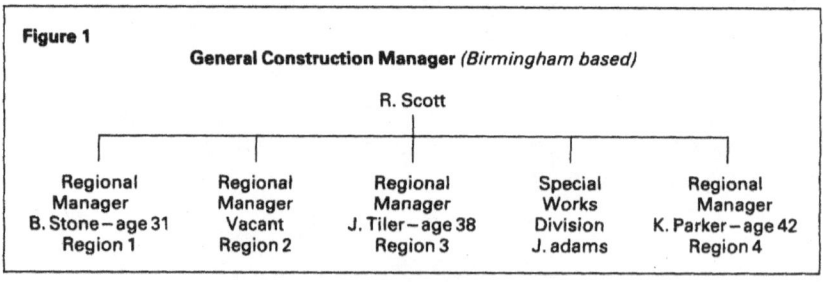

Scott finds himself with the following problems:
1. Two major contracts are so far behind the clients may not continue to do business with the company.
2. Techniques in Regions 3 and 4 are poor. There is no real cost control and no production control to any extent.

THE SCOTT CASE 33

3. There are quality problems on two contracts other than those mentioned in (1).
4. Profits are low. Turnover could fall and competition is intense.
5. He needs to fill his old job.
6. The image of the company is low.

Scott calls in Stone to help him to solve the problem which Stone does readily. this involves Stone encroaching in the other regions.

Stone who is very dynamic, very well-organised, intelligent and with a great capacity for work takes on more and more work. He is a married man with four young children and very ambitious. The contracts in his region show good profits.

He suggests to Scott that a design and build section ought to be established in Region 1 and that he would be prepared to manage it.

Scott comes to rely considerably on Stone's organising ability and is more than glad that the pressing problems of his new post are being dealt with.

There is a feeling of resentment building up amongst the other Regional Managers and they complain to Scott that Stone is handling too much work, in particular, work that Scott should be attending to. Tiler accuses stone of empire building. The morale of the regional Managers reaches an all time low.

Scott still feeling unsure of himself, says that he is merely exercising the management principle of delegation.

The value of work in the various areas is as follows:

Birmingham	£7 million
Region One	£13 million
Region Two	£12 million
Region Three	£18 million
Region Four	£11 million
Special Work	£4 million

The company does not operate in the SE.

Region 1 covers the Glasgow area as well as the North West of England.

Region 2 covers the South West part of the country including South Wales.

Region 3 covers the whole of the Midlands with the exception of Birmingham.

Region 4 covers South East Scotland as well as the NE of England.

Each Region has good office accommodation. The following gives the period of years with the Company for each executive.

SCOTT	20 years	Commenced with Company as a quantity Surveyor
STONE	4 years	Commenced as Area Manager
TILER	22 years	Commenced as Apprentice Joiner
PARKER	18 years	Commenced as Engineer
ADAMS	9 years	Commenced as Special Work Manager

The average profit before tax on turnover over the last five years is 2 per cent. The Head Office of the Construction Division is in Birmingham and occupies leasehold premises. The lease has four years to run. The organisation chart for the Head Office is shown in Appendix A. The organisation chart for Region One is shown in Appendix B, for Region Two in Appendix C, for Region Three in Appendix D and Region Four in Appendix E.

Regional Managers are responsible for forming their own organisation structure. A profile of each Region is shown in Appendix F.

The special work section under J. adams has been quite successful. Adams built it up from nothing. It involves all works under £100,000. The geographical area is approximately 60 mile radius of Birmingham. J. Adams has two estimators, one buyer, four contracts managers and on average 20 foremen.

The value of contracts carried out by the company vary between £300,000 and £8 million.

The net profit/loss before tax for the various areas last year was:

Birmingham	£150,000
Region One	£900,000
Region Two	£80,000
Region Three	£10,000 – loss
Region Four	£5,000 – loss
Special Work	£300,000

Appendix A – Head Office

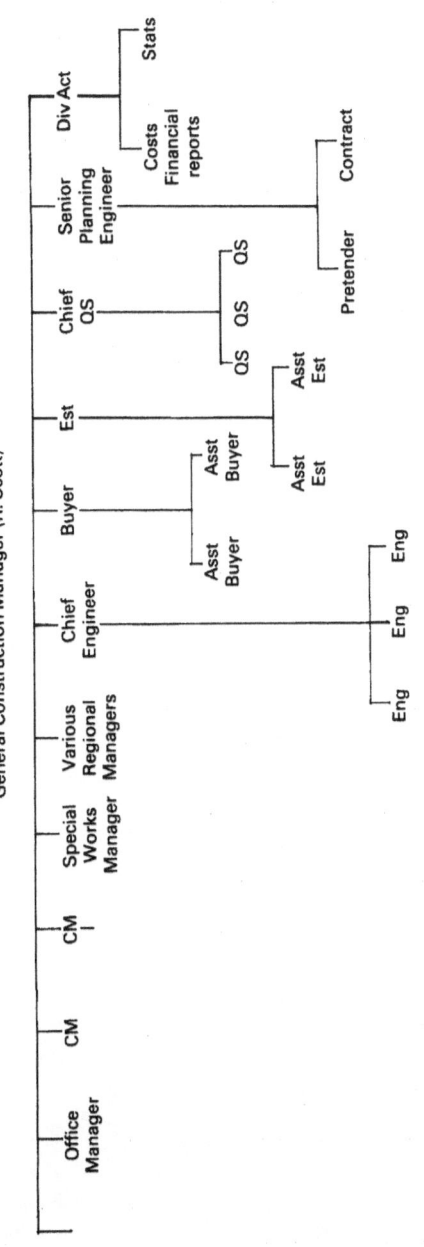

CM   Contracts Manager
Est   Estimator
Div Act   Division Accountant
QS   Quantity Surveyor

Appendix B – Region One

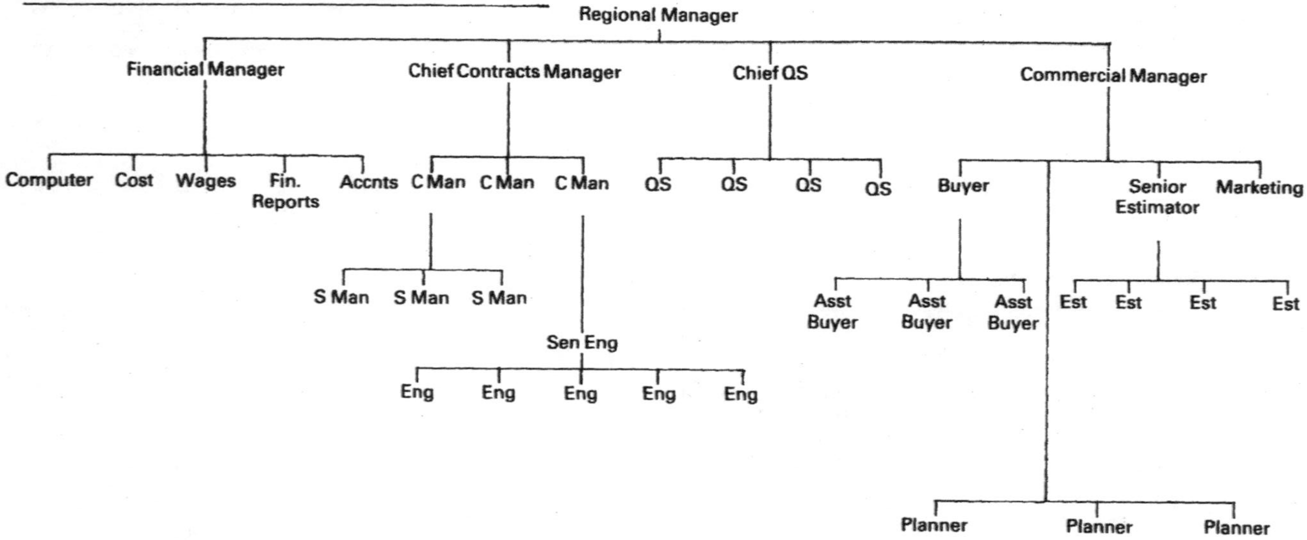

THE SCOTT CASE

C Man Contracts Manager
S Man Site Manager

THE SCOTT CASE

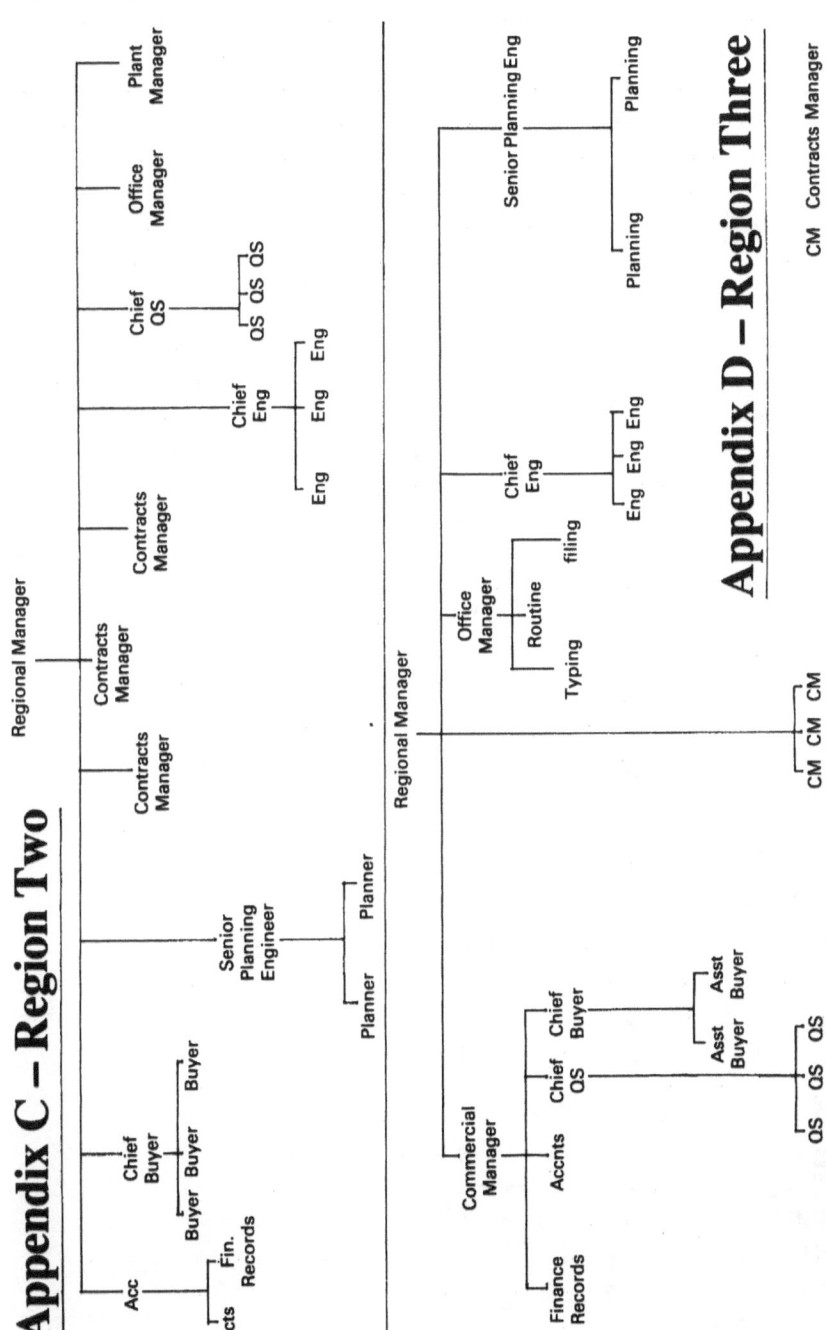

Appendix E – Region Four

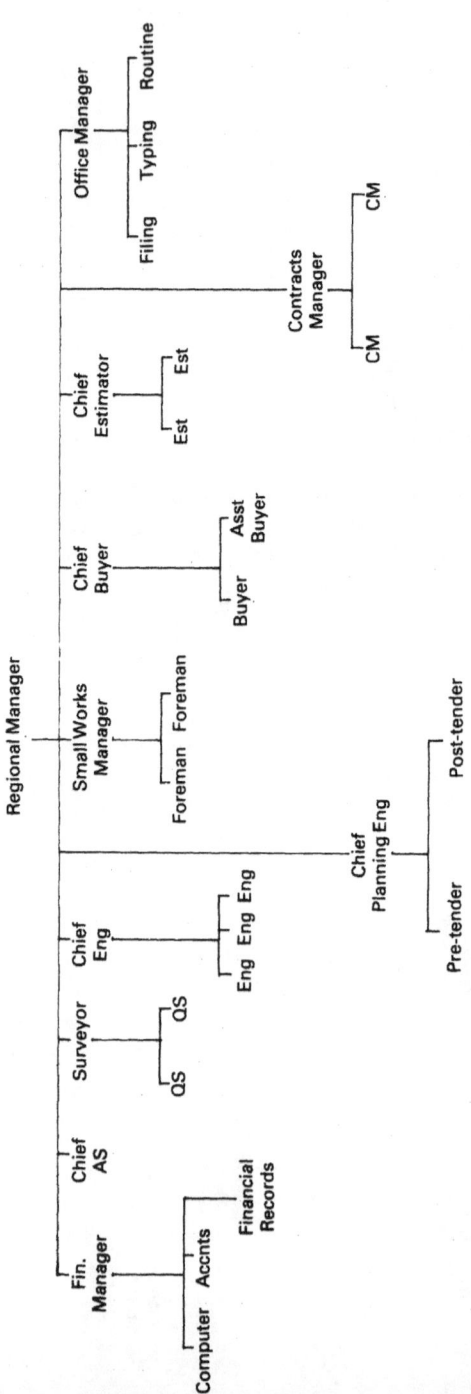

Appendix F
Profile of Region One

The majority of the work in this region is the 'one off' general building contract obtained in a highly competitive market. Stone believes in giving the site manager as much freedom as possible. He believes that he should also support the site and has adopted strict methods of control. For example, he insists that all sites send in a daily report sheet (see example in Appendix G) which he generally checks before he does anything else in the morning. If the site requires anything he makes sure they get it. Some people in Head Office consider he is too demanding but admire his drive.

Stone also insists that weekly programmes which support the long term programme are sent into Head Office for checking by himself. He is quick to put pressure on if the job is slipping behind. Stone insists that his contracts managers adopt the same methods of control especially when he is absent in other areas.

Stone is very keen to see that all staff are trained as much as possible; not only to do his own job but the job of the next step up.

A site managers/contract Managers meeting is held once per month.

The area is in the process of adopting a site information system based on the 'key word' idea. There was some resistance at first but it is now taking shape.

Stone is highly critical of the central organisation. He says they do not know what corporate planning is. The company does not appear to have any policy, objectives or corporate strategy.

Profile of Region Two

The work in this region has been dominated by timber framed housing and factory constructed units. The factory units arrive on site complete with the exception of the external brick skin. They are placed on prepared sub-structures. The idea was Scotts. It hasn't been a great success and the company are seriously considering reverting back to traditional methods. Scott had six sites working this method each with its own site manager. Scott left the general organisation to the site because he says he believes in delegating as much as possible.

Contract planning is left to a large degree to the site manager as long as he produced the production required. This, however, is rarely achieved. There is a lot of animosity building up between factory and site, each blaming each other for the ever increasing problems.

Profile of Region Three

Until recently a large part of the turnover of this region was in multi-storey flats for local authorities. The region was quite successful in this field but since the Ronan's Point disaster the orders have almost disappeared. The area then concentrated on large contracts in the £4.5 million plus region. These contracts are of a 'one off' nature. Tiler believes that an organisation that can build multi-storey flats can build anything.

The bulk of the work is sub-contracted on labour/material or labour only basis. Planning is very informal and is often on bar chart method. There is no formal short term planning. Tiler says it is almost impossible to do contract planning on this work as each job is a 'prototype'. He says it was easy on multi-storey flats but not so easy on these contracts.

Meetings are only carried out when absolutely necessary. Tiler thinks meetings are a waste of time.

Tiler believes in giving the site managers complete autonomy of their site and does not believe in interfering. Several site managers, however, say this is not the case in practice.

In the case of the large contract which is a long way behind programme, Tiler blames sub-contract labour, architects and the site manager.

Tiler is a very hard worker, a very strong personality and loyal to the company. He is firmly against any centralisation.

Profile of Region Four

The work in this region is divided equally between local authority housing, private housing, leisure building and 'one off' general building contracts. The Regional Manager blames the very keen competition in the area for the loss last year. He hopes to do better next year. The Regional Manager resents the intrusion of Stone into his region and objects strongly to the criticism from Stone. Stone says he is not using the right planning techniques his site procedures are poor, his cost control is non-existent and waste on site is deplorable.

Parker disagrees with this and says it is no worse than any other region or construction firm in the area. He does not believe in tight control and says site managers operate better if they are given freedom to interpret their role. He says, however, that good site managers are difficult to obtain. He doesn't believe in 'off the job' training for site managers as he claims it can only be done in a work environment. He says techniques such as network analysis and elemental trend analysis are not suitable for his contracts as each job is unique.

Parker believes in spending as much time on site as possible as he says this is where the money is made or lost. He doesn't like sitting in an office anyway. He is a hardworker commencing work at 8.00 am and never finishing before 6.00 pm. He will visit sites without informing either the contracts managers or the site managers.

The staff turnover in this area is particularly high but this does not worry Parker as he believes in injecting new blood into the company.

THE SCOTT CASE

Appendix G

Date								DAILY REPORT
J CONSTRUCTIONS LTD								Please send this copy to Head Office daily
P – Labour Present A – Labour Absent				RQD – Labour Required DTE – Date required				Please use reverse side if necessary
TRADE	P	A	RQD	DTE	TRADE	P	A	RQD	DTE	ORDERS PLACED
Bricklayers					Bricklayers					
APP										
Joiners					Joiners labs					
APP										
Painters										
APP										
Plasterers					Plasterers labs					
APP										
Plumbers					Plumbers mates					
APP										
Paviours					General labs					
Slaters										
Gangers										
Scaffolders										
Machine Op'rs										
Lorry drivers										
Timber men										WEATHER REPORT
Drain layers										
Kerb layers										
Steel fixers										
Steel erectors										
TOTAL					TOTAL					
Administrative	P	A	RQD	DTE	Sub Contractors	P	A	RQD	DTE	Gen details i.e. Vis to site, delays, stationery requirements, VOs Urges (in red) Acci, overtime & reasons
Agent										
General Foreman										
Asst to clerks										
Timekeepers										
Bonus clerks										
Storekeepers										
Cost clerks										
Works No.		Names of employees absent through sickness or industrial injury			Works No.					
										General Foremans Signature

Case 6
THE NATIONAL PLANT COMPANY

The National Plant Company specialises in the sale and hire of construction plant. They occupy some 35 depots throughout the UK and Southern Ireland. The Head Office is in the north east of England. They are a subsidiary of a large multi-national conglomerate whose activities range from construction to textiles, from food products to bricks. The turnover of the National Plant Company last year was £52 million and the net profit before tax was £2.6 million. The profit of the whole group was £96 million.

The Managing Director of National Plant would like to double turnover and increase profits by 300 per cent. He is convinced this can be done but feels that there is something wrong with the organisation which is preventing this. The additional finance to support the increase in turnover will be forthcoming from the holding company if a good case is put forward.

National Plant have increased their turnover and profit by 200 per cent in the last three years despite the recession and the Managing Director is convinced this could continue if he had the right type of organisation.

The present organisation is shown in Appendix A.

The board is made up of:

> Managing Director
> Operations Director
> One Member of Holding Co main board
> Four Regional Directors
> Technical Director (Associate Member)
> Company Secretary

The Managing Director, Mr Smith, has installed a computer system which will give quick feedback to depots on performance standards. The company operates on a strict budgetary control system. Management by objectives has not been formally adopted but some of the elements of MBO are used in the company.

Regional Directors are responsible for Area Managers but often bypass the Area Manager if they are visiting a depot. Area Managers don't like this and say that they don't always know what is going on.

The Regional Manager will often deal direct with top customers in the area and not allow the Area Manager to do so.

The average car mileage of a Regional Director is 70,000 miles per year.

The Irish Regional Manager is due to retire and has made some suggestions on re-organisation. He thinks that Area Managers are an unnecessary level which could be dispensed with. He is not sure what the Technical Director does and has received very little assistance from him. He claims to have only seen him at Board meetings. He does not feel he has been used to his full potential but cannot elaborate on this.

Salaries

	£
Managing Director	29,000 per annum
Operations Director	22,000 per annum
Regional Director	18,000 per annum
Technical Director	12,000 per annum
Commercial Manager	13,000 per annum
Financial Manager	13,000 per annum
Transport Manager	9,500 per annum
Administrator	9,000 per annum
Area Manager	12,500 per annum
Depot Manager	11,000 per annum
Foreman	7,000 per annum
Representative	7,000 per annum

The following is an outline of the executives in the Company:
The People
Mr Green *(Operations Director)*
Age 40, service with the company: 10 years.
Resident in Scotland.
Married with two children, aged 8 and 10.

Will not live in England as he prefers the Scottish Education system for his children.

Hard worker. Good knowledge of plant. Needs to spend a lot of time away from home as he travels the whole country.

He says he likes to keep his finger on the pulse in all *parts* of the country.

The results of a questionnaire are shown in Appendix C.

The results of a questionnaire to his subordinates are shown in Appendix D.

Further questions were asked and answers are as follows:

Q Do you delegate most of your work?
A I try to do as much of it myself. You can't always trust people to make the correct decision and if things go wrong, I will get the blame.
Q Do you carry out any management development of your subordinates?
A I do not get involved. We use a Lecturer/Consultant who gives my staff commercial training. I think he does a good job.
Q Do the Managers get an opportunity to practice what is taught?
A I haven't thought about it, but they have their jobs to do, so I suppose they must.
Q Do you see it as part of your job to develop management talent?
A No. They should develop themselves, after all, all management development is self development. No one developed me, I did it myself.
Q Would your subordinates like more responsibility?
A I haven't asked them. Most people try to avoid responsibility.
Q Would you like to see the Company expand?
A I don't think I could take on any more work.
Q Why have you got Area Managers?
A To take the day-to-day problems off my shoulders.
Q Would you like to become Managing Director?
A No.
Q Would you move to England to be near Head Office?
A No.
Q What are your main problems?
A Time and staff.
Q What do you mean by staff. Are the staff talented?
A Some are but don't show it. Some are not and don't want to be.
Q What changes if any would you like to see in the organisation?
A I think we should have a Sales Director who is also knowledgeable technically. This job is not like selling 'Mars Bars'. I don't find the Technical Director a great deal of help but he has a good knowledge of plant.
   I think we could manage with three Regional Directors.
Q Do you get good support from the rest of the Head Office staff?
A Yes. I often have disputes with credit control. They decide the 'cut-off' point with slow payers too quick. It is a highly competitive business and I don't want to lose customers.
Q Have you ever contributed to any changes in the Company apart from day-to-day events?
A No.
Q Do you like working for the Company?
A I suppose I must do. I spend most of my life here.
Q Do you manage to get through your work?

A No. I think there should be four of me.
Q Do you invite discussion on problems?
A I make the decisions and persuade the others to accept them.
Q Do they?
A They don't complain. It takes too much time to discuss everything.
Q Do you find your job stressful?
A Yes.
Q If you make a mistake does it really matter?
A I say it does. I operate with the yoke of the Profit and Loss Account around my neck. I can't afford to make many mistakes.
Q What does the Technical Director really do?
A He is supposed to give technical advice and attends to the 'nuts and bolts' side of training such as booking hotels. He does some product training himself. He travels around the areas. Some of the Regional Directors say he interferes too much in things he hasn't got anything to do with such as staff problems. He has a good knowledge of plant. He has a poor basic education, he cannot write a letter and cannot spell, but he has some good industrial contacts. He keeps a check on new products coming on to the market.
Q How do you get on with the Managing Director?
A Very well, he is a first class Manager. He leaves the 'nuts and bolts' side of the business to me. He is accountant trained so he doesn't get too involved in the technical side, although he likes to know what is going on and asks a lot of questions.
Q Have you received any training in delegation, business policy or financial management?
A I have been on six courses dealing with delegation, leadership and general management. I did not get a lot from them.

The following questions were put to the Regional Directors and the answers are typical of all the Regional Directors.
Q Do you have enough work to do?
A Yes, but I think I could take on more.
Q Do you get enough support on the 'nuts and bolts' side of the business?
A No. I would like a man in my Region. Problems are continually arising in operations management.
Q Do you think the Company could expand?
A Not with the present organisation.
Q Would you like to see it expand?
A Yes. It would create more opportunities for the talent we have in the Company.
Q Is there much management talent?
A Yes, a lot, but it has not been developed.

Q   Are you given a free hand to operate your region the way you like?
A   No. I haven't a great deal of authority to make decisions. I have on paper but not in practice. Most things go through the Operations Director.
Q   Would you like this authority?
A   Yes. I think it would develop me and in turn develop the people under me.
Q   What is your attitude towards delegation?
A   I think it is essential to the efficient running of an organisation.
Q   Do you ever by-pass your Area Managers?
A   Sometimes I have to as the Depot Manager may not be in the Depot.
Q   Why not?
A   He is expected to spend a lot of his time outside the Depot obtaining business.
Q   I thought that is what the Rep did?
A   They both do it.
Q   Who is in charge of the Depot when the Manager is out?
A   The Foreman looks after the workshop. The hire Clerk looks after the office side.
Q   Has each Depot got a hire Clerk? Is she/he trained in any way?
A   Yes and they receive no formal training although they are very often dealing direct with the customer.
Q   Could you do without your Area Managers?
A   I could, but some Regional Directors would like them to remain.
Q   Why?
A   It removes some of the load from them.
Q   Do you have meetings with the Area Manager and Depot Manager?
A   Yes, once every three months. We discuss common problems and I give information. The Minutes are sent to the MD after I have checked them.
Q   What do you mean by 'checked them'.
A   Edited them.
Q   So the MD only reads what you want him to read.
A   More or less.
Q   Do you involve your subordinates in decision making and problem solving?
A   Yes, but I make the ultimate decision – the buck stops with me.
Q   Are you happy with all your Depot Managers?
A   No – there are one or two I would like to get rid of but the Operations Director will not let me.
Q   Do you ever raise these points at the Board Meeting?
A   Not really, We are very limited to what we contribute to the Board Meetings. The Main Board Director and the MD dominate the meetings.

Result of Interview with the Depot Manager

Q Do you have enough authority to carry out your work?
A No. I can't make as many decisions as I would like.
Q Could you elaborate on this?
A I can't decide on the type of plant and the number of units. I can't dismiss or employ without the Area Manager giving me special authority. I have been trained in industrial relations legislation but get little opportunity to use it.
Q Do you like the present arrangement of being out of the Depot a lot?
A I can't see any other way it can be done. A lot depends on how good the Foreman is. I do need to visit customers and potential customers.
Q Do you get a lot from the meeting with the Regional Director?
A I don't think they are much good. He does not make any progress. The Minutes to the MD are fixed. It is very difficult to make direct contact with the MD. I would like to see more of him in the Depot.
Q Does the Area Manager give you a lot of support?
A No. It is more interference. I could really manage without him.
Q Why does the Company have Area Managers?
A I am not *sure*.

Meeting with Area Manager

Q Do you think the Company could do without Area Managers?
A No.
Q Why not.
A The geographical spread is too great. Regional Directors would not be able to cope with the supervision of Depot Managers.
Q Are the Depot Managers issued with budgets.
A Yes – annual, monthly and weekly.
Q If the Depot Manager was not keeping up to target, would the Regional Manager quickly know.
A Yes, very quickly – not later than one week.
Q What do you actually do?
A Supervise Depot Managers, visit client with Depot Managers, produce reports, order plant, attend meetings. Employ and dismiss operatives and other things in general.
Q Do you think the Company should expand?
A Yes, the potential is great.
Q Would you like promotion?
A Yes.
Q Does the Company have a management development programme.
A No. I also think they should carry out appraisal of staff.
Q Would you like to do it?
A No.
Q Have you a great deal of authority?

A I would say, very little.
Q Why?
A The Operations Director involves himself too much in details.
Q What is your opinion of the Operations Director as a Manager?
A I would rather not comment.
Q What about the Regional Directors?
A One is retiring soon, two are good, but the one I have is poor. He talks a lot and does little. I believe in action rather than words.

Profile of Remaining Executives:
Commercial Manager
Good basic education. BSc in economics. Age 30.

Training and experience principally in administration. Technical and accounting knowledge poor. Knowledge in the areas of personnel and training is very good.

Sales knowledge and experience is very limited.

He is a good general organiser with an excellent personality. Good motivator of personnel. He is willing and quick to learn.

Married with two children.

Financial Manager
Age 45, bachelor. Qualified Accountant. Very thorough, will not make a decision unless he has all the facts in front of him. Dull personality. Excellent knowledge of Financial Management. Very hard worker. Invariably works late. Good organising ability. Has some good commercial ideas.

Technical Director
Age 50. Married, no children. Poor basic education.

Good knowledge of plant, served Apprenticeship as a fitter and became Salesman. He had a lot of success as a Salesman but would have difficulty controlling a sales force.

Transport Manager
Good education. Qualified Transport Manager. Very thorough. He is not too well liked by the line Management as he makes a lot of demands on them.

Age 35. Married with two children.

Administrator
Good basic education. Commenced with the Company when he left school at 14, 20 years ago. Good organiser, thorough worker. No recognised leadership ability. Recognised to be good at his job. Married, no children.

Mr Smith will be promoted to the main board within the next two years and he would like to carry out any re-organisation before that. He will also need to nominate his successor.

At main board level he will be the link between National Plant and a further six Companies in the Construction division and the main board.

THE NATIONAL PLANT CO LTD CASE

Appendix A

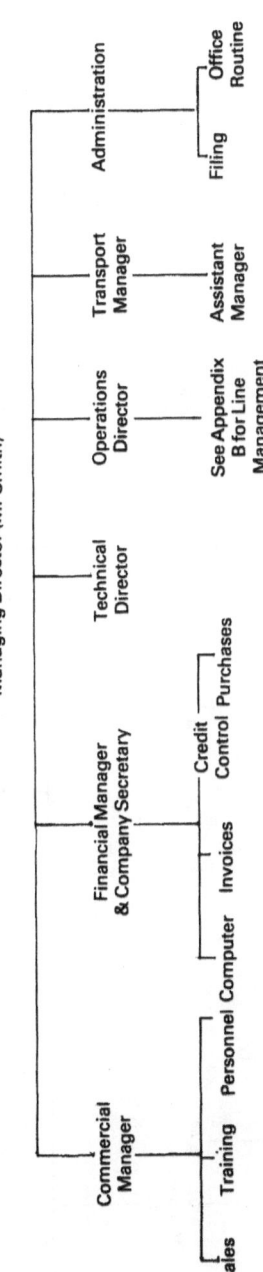

Appendix B

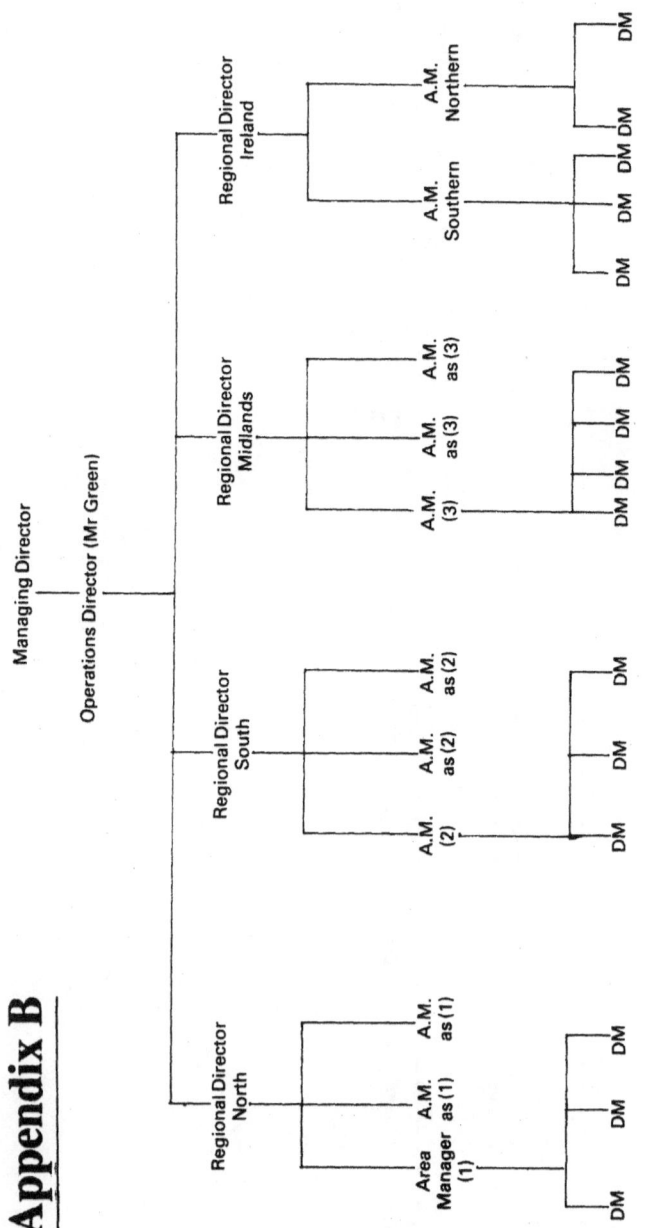

Appendix C

Answers from Operations Director, Mr Green.
1. Are unexpected emergencies constantly arising in the operations you supervise?
   A   Yes, but that is inevitable in this business.
2. Do you find that your daily job consumes so much of your time that you have no time to plan?
   A   I plan as I drive in my car.
3. Do you get bogged down in detail?
   A   Sometimes, but I need to. I am the only one to sort it out.
4. Is there a lot of friction or dissatisfaction in your work group?
   A   Not a lot.
5. Do simple jobs take a long time to get done?
   A   Unfortunately, yes.
6. Do you complain or criticise others when the work you are responsible for does not go as planned?
   A   I tell them in no uncertain terms if they do not produce the goods.
7. Do your subordinates always wait for you to give the sign before they start a job?
   A   Generally, yes.
8. Have subordinates stopped coming to get reactions to their ideas?
   A   They do not produce many ideas.
9. Do your employees display little or no enthusiasm in their work?
   A   They are generally enthusiastic.
10. Do you believe you should take some risks?
    A   Yes.
11. Do you feel you could do the work of your subordinate better yourself?
    A   Yes.
12. Would you quickly know if things were going wrong in the work you supervise?
    A   Not always because of the distance involved but performance returns produced by the computer are good.
13. Are you receptive of other people's ideas?
    A   I suppose so.
14. Do you encourage subordinates in their work?
    A   I try to, but I expect them to do their job. That is what they are paid for.
15. Do you have a lot of unfinished jobs accumulating?
    A   Yes.

Appendix D

Answers from Regional Director (this was typical of other Regional Directors).
1. Do you have a clear understanding of your duties and responsibilities?
   A   More or less.
2. Do you, generally speaking, have the facilities and opportunity to carry out your tasks?
   A   Not always.
3. Does your superior get involved in detail of your job?
   A   Yes.
4. Does your superior get involved in results?
   A   Yes.
5. Is there an easy flow of communication between your superior and you.
   A   I find communication difficult because of the distances we are spread over.
6. If you perform well will you be rewarded in some way?
   A   I do not think so.
7. Are you kept informed on the progress of the work for which you are responsible?
   A   Yes – through the computer system.
8. Are you allowed to speak freely in the organisation?
   A   Sometimes, not often.
9. Does your superior consult you on problems you are familiar with?
   A   No.
10. Are your contacts with other parts of the organisation based on status?
    A   Yes.
11. Are your contacts with other parts of the organisation based on tasks?
    A   No.
12. Are the controls for your job satisfactory?
    A   Yes, to some extent but I get frustrated because of lack of support.
13. Do you receive instructions direct from your immediate superior and not two/three levels up?
    A   Sometimes.
14. Are you expected to know all the details of your subordinates at all times.
    A   Yes.
15. Do you like responsibility?
    A   Yes. I would like more.
16. Do you fear criticism from your superior.
    A   Yes.

Case 7
NORTHERN CONSTRUCTION LTD

The Northern Construction Company operates North of the Humber and South of the Scottish border. It carries out General Building and Civil Engineering work with a labour force of 1,500 men plus 150 office staff. The annual turnover is approximately £75 million.

The following cases have occurred within the Company. How should they be handled.

Case 1
The Teesside school site is having trouble with a squad of bricklayers (4 + 2) coming in late each morning by approximately 30 minutes, the site manager has remonstrated with them but they say it doesn't make any difference to the firm as they are on bonus and it is, therefore, their loss. The site manager feels that their action is affecting the morale on the site.

Case 2
J. Smith a joiner on the city centre site, is generally speaking a poor joiner. He is on the slow side and the quality of work is not up to the company's normal standard. All the joiners refuse to have him in their team. You have warned him to improve without result and so you terminate his employment. Smith is a member of UCATT. The card steward calls a meeting of the men who, on a show of hands, pass a resolution that if Smith is not reinstated the site will be on strike.

Case 3
Northern construction want to employ a new Quantity Surveyor and Site Agent for the general building work. Draw up a job description and explain how recruitment should take place. Write up outline questions to ask at the interview and draw up an induction programme.

Case 4
The GF on the hospital site at Durham appointed a chargehand from a

squad of bricklayers (4 + 2) the bricklayer, J. Brown, is certainly the best bricklayer in the squad, hence the reason for his appointment. The morale of the squad immediately falls and is in danger of breaking up. J. Brown appears to be a reasonable man and no one in the squad has made any complaints about him. He is the oldest member of the squad with a lot of experience. What are the possible causes of this?

Case 5
A hod carrier working on the Teesside site, does not turn in for work on Monday morning, this upsets the balance of the squad. The GF instructs a member of the concreting squad to take his place. He refuses, the GF warns him and then sacks him for refusing to obey an order. He then instructs a second member of the squad and the same thing happens. He eventually sacks the whole of the concreting squad (five men) the concreting squad is a good group and the loss will be felt by the site.

Case 6
Absenteeism in the plastering squads is very poor (average 1½ days per week). Plasterers are very difficult to get in the area and the contract is running behind programme.

Case 7
J. White has worked for Nothern Construction for 20 years. He is a good loyal site manager. Recently he is showing all the signs of alcoholism. What action would you take?

Case 8
You have been requested to carry out a training survey of Northern Construction Limited. How would this be done?

Case 9
The MD of Northern Construction is not satisfied with the quality of bricklayers being produced by the company, the difficulty appears to be with bonus squads. Unwillingness to accept the apprentices into the squad in case their bonus earnings should drop. What action should be taken?

Case 10
A new cost control system has been installed but is in danger of collapse as the site managers and plant operators often fail to complete returns. The MD cannot understand this as the last company he worked for used a similar system without any trouble at all.

Case 11
The scaffolders on the city centre site walked off the job last Friday lunch

time saying they intended to hold a union meeting, this caused some of the work to be disrupted. The site manager warned them about this but the same thing happened the following Friday, the scaffolders work should take about six weeks to complete from now. What action would you take?

Case 12
The GF notices one of the plumbers putting a brass fitting into his haversack. He approaches the plumber and accuses him of this, the plumber denies having the fitting so the foreman asks him to open his bag. The plumber refuses saying that the foreman has no right to do this and if he attempts to open the bag he will sue him for assault and battery. What action should be taken?

Case 13
The General Foreman on a large construction site informs you that he is having great difficulty getting from his home by public transport. He has been offerd another job with a company car and says he intends to take this if the firm do not make a similar offer. The foreman is a first class man and may be very difficult to replace.

Case 14
The plasterers are making a claim for an extra £120.00 bonus which they say they have lost because of hold up by a sub-contractor, the sub-contractor is nominated and has worked on the contract all the time he was allocated for his work.

Case 15
George Steel is a 35 year old contracts manager with a DMS and degree in Building. He has an exceptionally good record with the firm and could one day be a top line executive. You have heard on the 'grape vine' that he is looking for another job because of his lack of progress in Northern Construction. There are no vacancies for executives above George's level at present and there are unlikely to be any for the next four years, (except in the case of death). As MD you will be very sad to lose Steel as he has great potential. What would you do?

Case 16
Most sites are experiencing difficulties in recruitment caused by the fact that when jobs are advertised, men will phone up and say they will appear on Monday morning. Further telephone enquiries after that are told that the job is filled and on Monday morning the men do not turn up.

Case 17
You walk into the joiners shop during work hours and find one of your

oldest men doing a private job. What action would you take?

Case 18
When applying for labourers from the Department of employment it is quite common to receive several men who produce a green card and ask you to mark it unsatisfactory. This enables them to return to the dole without loss of benefit. What action would you take?

Case 19
The shopping site at Hexham is suffering with very low morale. This isn't usual for Northern Construction who prides itself on good men management. The low morale shows by high labour turnover on the site, low productivity and several minor stoppages. You have been requested to investigate. What points should be looked for and what should be done about it. The conditions on the Hexham site are very good. Most of the labour used is local labour, the job is not particularly hazardous.

Case 20
Two of your senior surveyors have approached you independently complaining that the other is getting on his nerves. When asked in what way and why, each man has difficulty in giving answers. The position has got so bad that one or both may leave the company. What could be the cause and what should be done?

Case 21
The men on the Tynedale housing site have demanded a canteen on site, the number of operatives at peak will be 40. Would you comply with their demand.

Case 22
The foreman on the Otterburn site has requested another concrete mixer with gang on a job which requires high quality concrete. The site already has one gang of men who have worked together for six years. They are first class and can be relied upon. A mixer is delivered to the site and a new gang recruited. To ensure quality one of the men, Bill Smith, is taken from the first gang and told to lead the new gang and will receive an extra 6p an hour. He does not complain about this but is obviously far from happy. Three days later the first gang request a meeting and say that if you do not sack the labourer who replaced Bill Smith they will all leave. They say he is no good, although you have employed this man some two years ago and feel that he is a good man. What could be the reason for this action by the men?

Case 23
John Hall has complained that he is being victimised on his site by the

foreman. He claims the foreman gives him all the bad jobs and says he does this because he is a member of the National Front and the foreman is a communist. John Hall is a joiner and has been with the company two years. He is a good joiner as far as you know. Comment on this case.

Case 24
J. Rigg is a Drott driver and is a very good one too. His machine broke down through no fault of his own and the foreman puts him on general labouring until the machine is repaired. J. Rigg resents this and shows this by low productivity. Comments on this case.

Case 25
There is a demarcation dispute on the city centre job because the members of the AEU say the job of fixing wood beams to steel beams with steel bolts is their job. There is a risk that the job could be brought to a standstill. What are the psychological aspects of a dispute of this nature?

Case 26
The safety record in the company is far below standard. All the men have attended safety courses, the safety officers bombard the men with propaganda but still the accidents occur, the MD thinks the psychological approach should be adopted. Could it? and if so how?

Case 27
T. Murray is a good labourer and is particularly good for trowelling floors, concrete bays and the like. The UCATT union has said this is the work of a bricklayer, plasterer or paviour and Murray should stop doing it. Have the union got a case?

Case 28
At the joint consultative meeting within the firm the men's representatives have requested that all time clocks be removed from sites as the Company should trust its men. Would you comply with their request?

Case 29
ABC (Painters) Ltd having had an unfortunate experience with a painter who had got drunk during the lunch break and did £2,000 of damage on a contract, made it a rule that no-one would be allowed to drink intoxicating liquor during the lunch break.

Three painters, Smith, Brown and Green were discovered coming out of a local at 1.00 pm by the contacts manager. He issued a verbal warning to the painters with the foreman as witness. Two weeks later he discovered the same men plus John White coming out of the same pub. He then issued a written warning to the four men.

Three weeks later he was told they had gone to the pub for lunch. He visited the pub and found all four drinking. He then dismissed them.

White brought a case of unfair dismissal because he claimed that, unlike the others, he had not been given a verbal warning.

Did White have a case?

Case 30

ABC (Painters) Ltd are a subsidiary of Northern Construction. The Company did a great variety of work ranging from bridges painting to re-paints of fine buildings such as theatres, high class hotels and museums.

The Darlington depot specialised in the high class work whilst the structural side was carried out from Newcastle. Owing to the present economic climate the amount of high class work had reduced considerably although the sturctural side is still very busy.

John White and Bill Green were both good craftsmen they had been employed at Darlington for some four years.

ABC Ltd instructed White and Green to report to the Newcastle depot in future as there was insufficient work at Darlington. White and Green objected to this on the grounds that the work at Newcastle is inferior and it meant a 25 mile journey from their homes.

ABC Ltd agreed to pay travelling expenses from their home. White and Green agreed, though very reluctantly to go. Four weeks later they left the firm and claimed redundancy pay. Did they have a case?

Case 31

You receive reports from contracts and it is noted that on Contract 56 a school for the county council, work is not proceeding as smoothly as could be expected. Complaints have been received from the county Architect about bad workmanship, particularly brickwork. Contract costs are high and work is behind programme.

The GF on the job also acts as foreman bricklayers with two chargehands to assist. Other trades have their own respective foremen. The work is valued at £1,950,000.

On checking the GF's record, you find he has been with the firm for 12 years. He has a good record of loyalty and is trustworthy, punctual, has no fear of the men and does not hesitate to speak his mind. He is reluctant to adopt new methods and offers resistance to change. He does not possess high intelligence and does no thave great drive. The job is run on a bonus scheme. Labour is similar to that existing in Newcastle.

What action would yopu take and why?

Case 32

Mary Jones, who has been your clerk for four years, informs you that she is pregnant and will be eventually leaving the job and says she would like to return. Fourteen weeks before the baby is due she hands in her notice to terminate her employment because of sickness and informs you that she will be returning to work before the end of the 29th week after the baby is born.

Eleven weeks before the baby is due she sends in a claim for maternity pay. You have employed a new clerk called Betty Smith.

Twenty-eight weeks after Mary's baby arrived she sends a letter saying she would like her job back. You don't want to terminate Betty's employment so you try to get Mary another job at another office within the firm which is within travelling distance of her home. The job will be similar to the job she did at your depot. You succeed in doing this but Mary says she wants her own job back.

You give Betty two weeks notice and tell Mary to start work again. Betty then sues you for unfair dismissal.

On the Thursday of the 28th week Mary says she will be unable to start for a further two weeks because of sickness. When she eventually starts she informs you that she is pregnant again.

Comment on the above case.

Case 33

Jim Godwin is a joiner at your yard. He is generally speaking, a good worker, always punctual, always there when he is required. Jim can best be described as a 'loner'. He treats his fellow workers with a degree of contempt and knows he is good at his job and lets his fellow workers know also.

His colleagues tolerate him, but don't fully accept him. They will not co-operate with him if they can possibly avoid it. When Jim is around there is always a degree of tension in the depot.

He complains a lot to his foreman, sometimes over the most trivial things. No one likes to work with him and he doesn't like to work with anyone else.

You have checked his background and find that he was foreman joiner in the joiners shop of a large National Building Contractor. He says he left the job because the men would neither work hard enough or good enough. His wife left him five years ago with two children, both of whom are now married and live away from home. He lives on a large council estate on his own and is now 48 years old.

What would you do with *Jim?*

Case 34

Bill Short was a difficult man to work with. He upset everyone on the site. No matter what anyone said Short would disagree. He accused everyone on the site of almost every sin in the book. He was, however, a reasonable worker and generally obeyed instructions given by the foreman, but invariably made some sarcastic remark on receiving them.

Eventually Short's behaviour got on top of the men and they all approached the foreman on Monday evening after work and said that they would no longer work with Short. If he was not dismissed they would all be out on strike next Monday.

Short was told of this and greeted this with the usual contempt he had for his fellow workers. The Manager tried to arrange a transfer but failed. The manager and the foreman had spoken to Short about his behaviour on several occasions but they both formed the opinion that he would never change. The manager decided that Short should be dismissed for misconduct.

Short took the case to the tribunal for unfair dismissal. Do you think he has a case?

Case 35

Bill Jones, lorry driver, had been repeatedly told about the securing of equipment on his lorry but was inclined to ignore the advice. He was generally, otherwise, a good worker.

He was delivering a large pump to a construction site one day when it fell off the wagon doing £500 worth of damage. On instruction from the manager the foreman gave the driver one weeks notice. He finished work right away and went home. Later he was paid for that week and also given two weeks pay in lieu of notice.

He applied to the tribunal on the grounds of unfair dismissal. Did he have a case?

Case 8
NORTH EAST DEVELOPMENTS LTD

The Southern Securities Finance Company have purchased 600 acres of building land ripe for development in the North East. They propose to build houses in the lower bracket on it (£27,000–£41,000). A subsidiary has been set up and you have been appointed Managing Director. Your appointment commenced last Monday and the company expect a start to be made in seven weeks' time. You will be starting from scratch and you have been give a free hand provided you show results. Planning approval has been obtained for 500 units and outline planning approval for 5,500 units. Units must not exceed two storeys unless a re-submission for approval is made.

The drawings for the 500 units are completed, but you are responsible for the production of further drawings required in the future.

The houses will be sold freehold.

Site works have not commenced.

You have no labour, plant or materials.

You have rented accommodation in Newcastle consisting of seven offices plus ancilliary accommodation.

You have no yard.

You are being paid a generous salary plus 5 per cent of the profits.

This is not a 'once and for all' project as the company would like you to build up a continuing organisation in development work of all kinds. Your future will not be limited to the North East if you so wish.

The syndicate should produce a report dealing with the:
a. Procedures for getting off the ground;
b. Organisation;
c. Objectives;
d. Policy;
e. Strategy;
f. Recruitment;
g. Planning;

h. All the functions required for an organisation of this nature;
i. Finance required.

Initial overdraft facilities of £100,000 have been arranged at your local bank. Finance may be obtained from the parent company provided you can show that a good return will be forthcoming.

Case 9
DISCIPLINE CASES

Comment on the following and state what action you would take.
1. A Joiner has been employed by the company for six years. He is a very good joiner and a high bonus earner. He is very willing, pleasant and helpful but he takes, on average, three half days off each week. You have told him about it on three occasions but there is no improvement. There is, to the best of your knowledge, no good reason for this.
2. A Joiner who has worked for the company for 10 years is caught stealing 3 No cylinder latches. What action would you take? The rest of the men on the site know he has been caught. Will it make any difference if he had only worked for the company for six months.
3. You have on several occasions found your clerk making long expensive telephone calls.
4. One of the labourers has threatened the foreman on two occasions with violence. He has a record of violence outside the company.
5. One of your joiners produces a very good job but takes a long time to do it. He never makes bonus, apart from the guaranteed bonus and often loses on the job.
6. A mixer operator often fails to clean down his machine and does not check water and oil levels on a regular basis.
7. A full pallet of bricks is placed on the scaffold although the bricklayers only need two dozen to finish off the lift. The scaffolder tips them off when he is about to lift the scaffold.
8. A bricklaying squad shouts for a tray of mortar at 4.15 pm with the result most of it is left when they leave the site.
9. A labourer is working in a 3 m deep trench without trench timbering although this has been provided. He tries to re-assure you that the ground is firm and does not need it.
10. A fork lift truck driver is supplying three gangs of bricklayers. two of the gangs complain that the other gang received preferential treatment with the result they often have to wait. The gang and the driver

deny this and say it is just an excuse because the other gangs cannot produce as much.
11. You have, on several occasions, found a young joiner bullying a young apprentice. You have spoken to him about this and he says it is just having a bit of fun and the bullying continues.
12. You have found a joiner for the second time in six weeks cutting a 1 m length off a joist which was purchased the correct length for a floor.
13. You have discovered a labourer causing a terrible mess in the toilets.
14. A dumper driver has been caught carrying a passenger for the third time in a week.
15. Bill Jones is generally speaking a good joiner but makes more than the average number of mistakes hence wasting a lot of materials.
16. One of the operatives on your site has had six accidents of relatively minor nature in the last four weeks. He has had four serious accidents since he commenced working for the company four years ago.
17. You have discovered a bricklaying squad removing a guard rail and failing to refix it. This is the third time in two weeks.
18. A ganger has been absent on 10 occasions during the last eight months ranging in periods of three days to one week. His sick note usually says cold or bad back.
19. One of your men is a general trouble maker, grumbles and complains about almost everything and encourages the rest of the men in the site to do the same. Morale on the site is low and this could be one of the causes.

Case 10
THE 4½ DAY WEEK CASE

You are manager of CN Construction Ltd, employing approximately 300 men consisting of Bricklayers, Joiners, Painters and Labourers. About three-quarters of the sales turnover is in private housing.

In March of this year it was noticed that the great majority of the men employed on the housing sites finished work at Friday noon, having worked a 4½ day week. The idea spread to other sections of the Company and you were approached by a recognised shop steward to put the whole of the organisation on a 4½ day week during the summer months. The representatives proposed that the way to do this was to work extra hours on Monday to Thursday to make up the 39 hour week. Non productive time would not be paid for these extra hours as they were for the convenience of the men.

The idea appealed to you because you still had to employ site staff for five days in case some of the men worked. If a 4½ day week was adopted everyone would know exactly how he stood. You agree to the idea and everything worked well until you employed John Smith of UCATT on 7th June. Smith was a very strong union man and had only worked for the Company for two weeks before he argued that all the men should be paid for the non-productive part of the overtime they had worked since 16th March to make up the half day lost on the Friday. He reported the matter to the Regional Secretary and all the unions who had workers on your site sent in a joint claim for £4,000 back money for the non-payment of the NPO.

Have the Unions got a case? If you feel that they have not prepare your case for submission to the Local Joint Council.

Case 11
PROPOSED CONSORTIUM – CASE STUDY

You have been requested by three building companies to investigate the possibility of, and make recommendations for, the formation of a consortium of the three comanies.

Each company is willing, if necessary, to lose its identity in a new company. Alternative arrangements will be considered.

The reason why the companies feel that they should merge in some way is because of the infiltration of large national contractors into the area taking on work which they, as individual companies could not tender for. They also feel that their resources could be better employed if a grouping should take place.

The main shareholders and managing director of each company are very good friends and meet regularly on a social basis.

Details of the three companies are attached.

TYNE CONSTRUCTION COMPANY LTD

Mr Green is Managing Director and holds 80 per cent of the shares. His wife holds 10 per cent and a brother, who does not take an active part in the company, holds the remaining 10 per cent.

The company employs 30 men, made up of labourers, plant operators, bricklayers and joiners. They sub-let all other works.

PROPOSED CONSORTIUM CASE

Account balance sheet is as under:

TYNE CONSTRUCTION COMPANY LTD
Balance Sheet as at 30th June 19......

Authorised Capital 100 – £1 shares

Issued and fully paid up capital		Freehold land	10,000
100 – £1 ordinary shares	100	Plant less depreciation	28,000
10% loans to company		Land for development	30,000
Green	60,000		
Green's brother	28,000	Sundry debtors	30,000
		Work in progress	
Secured Loan	20,000	less payments on A/C	69,000
Future tax	15,000	Stock	2,100
Current tax	16,000	Cash in hand	10,000
Sundry creditors	40,000		
	£179,100		£179,100

The secured loan is using Greens' own house as colateral; the articles of association are 'off the peg'.

Tyne Construction Co Ltd is situated in the heart of the town. assume book value of plant is correct. The turnover last year of Tyne was £700,000 and this year it should be £600,000 for 12 months up to the end of June. The profit last year was £40,000 and this year it should be in the region of £25,000. Assume that the date is 1st August. the freehold land is used for the company's yard and office. Planning permission has been granted for six houses on the land for development. The figure in the balance sheet is at cost. The land was purchased six months ago.

The company has the following work in for this year and next year:

1 school	£100,000 due for completion next January
1 hospital (small)	£250,000 due for completion next March
1 child welfare centre	£190,000 due for completion next April
60 council houses	£1,800,000 due for completion next June

The company employs:

	Salary
2 general foremen	£8,000 each
2 non productive trades foremen	£7,000 each
1 storekeeper	£4,500
1 estimator/QS	£9,000
1 contracts manager (Green)	See below
2 office clerk/typists	average £4,000 each

The plant consists of:

2	30 cwt hoists	8	5/3½ mixer
2	5 cwt hoists		Small tools
2	8 tonne tippers	1	porto-saw
8	Dumpers (15 cwt)	2	JCB 3 C's
1	21/14 mixer	5	company cars
1	18/12 mixer		Site huts
2	14/10 mixer		Scaffold
3	10/7 mixer		

They buy in all their joinery; They have a bonus scheme in operation; They use bar charts for planning; Green's salary last year was £17,000; The company was commenced by Green in 1974; Overdraft facilities = £15,000.

THE BROADWAY CONSTRUCTION COMPANY LIMITED

Commenced business in 1932 by Walter Smith's father. Smith's father died in 1948 and Walter inherited the business. His old mother has money invested in the company at an interest rate of 10 per cent amounting to £10,000. It is expected that Walter will inherit this together with property valued at £50,000 when she dies as he is the only son.

He himself has two sons who are in the business –

Jim Smith is 30 and is a contracts manager. He has a BSc Building (Manchester).

Joe Smith is 28, BSc (Herriot Watt) in Quantity Surveying. He is now the chief QS.

The company employs 200 men consisting of labourers, plant operators, bricklayers, joiners, plumbers and two electricians. They sublet the rest of the work. The turnover last year was £3 million and this year should reach £3.5 million. the profit for this year should reach £170,000. Last year the profit was £230,000. The firm has a bonus scheme in operation and uses sophisticated planning techniques. Jim Smith has always wanted to use work study but his father has objected to this on the grounds that the company could not carry a full time work study man.

PROPOSED CONSORTIUM CASE

The staff of the company consist of:

		Salary
Chairman	Walter Smith	£21,000
Managing Director	G. Pattison	£17,000
Chief Contracts Manager	Jim Smith	£17,000
Chief QS	Joe Smith	£17,000
2 contracts managers		£16,200 each
6 General foremen		£11,500 each
4 Site agents		£13,000 each
3 full time trades foremen		£11,000 each
4 Site clerks		£7,500 each
1 Office manager		£11,800
2 Bonus clerks		£7,800 each
2 Estimators		£11,800 each
1 Buyer/Surveyor		£11,000
2 Surveyors		£11,800 each
6 Office girls	average	£7,000 each
2 planners		£6,000 each

95 per cent of the shares are held by Walter Smith.
5 per cent are held by his wife.

The contracts on at present (work to be completed) are:

		Due for completion
1 Primary School	£1,300,000	next April
1 Comprehensive	£1,450,000	May
80 LA Houses	£2,330,000	June
20 Private Houses	£1,600,000	estimated November
1 Aged persons home	£430,000	next June
1 Child Welfare Centre	£190,000	March
1 Health Centre	£820,000	February
1 Supermarket with flats	£850,000	March
1 Hospital	£2,500,000	June
1 Working Mens Club	£420,000	January

THE BROADWAY CONSTRUCTION COMPANY LTD
Balance Sheet as at 30th June 19......

Authorised Capital £2,000 in £1 shares

Issued and fully paid up		Freehold land	20,000
2,000 – £1 ordinary shares	2,000	Plant less depreciation	30,000
Capital reserve:		Debtors	50,000
Revaluation Reserve	10,000		
Revenue reserve	140,000	Work in progress less payments on A/C	253,500
Loans unsecured	80,000	Stock	1,500
secured	10,000		
Future tax	35,000		
Current tax	30,000		
Sundry creditors	40,000		
Bank overdraft	8,000		
	£335,000		£335,000

The plant consists of:

1	40 cwt hoist (skip)	1	14/10 mixers
2	30 cwt hoists	5	5/3% mixers
4	10 cwt mobile hoists	6	15 cwt dumpers
1	5 cwt scaffold hoists	7	company cars
1	Hymac		Site huts
1	JCB 3C		All other plant is hired
2	21/14 mixers		

THE TYNEMOUTH CONSTRUCTION COMPANY

This company has been in business since 1950. It employs 40 men including labourers, bricklayers, joiners, plasterers and painters. They have a well equipped joiners shop and employ four joiners on a regular basis. The company has a good yard site and owns six acres of adjacent land in the suburbs. It is managed by John Rose who has 60 per cent of the shares, the remaining shares being divided equally between two sisters. There is an article indicating that the business cannot be sold except by extraordinary resolution.

John Rose has no family. He is 56 years old.

PROPOSED CONSORTIUM CASE

The business was purchased as a going concern in 1950. Overdraft facilities exist for £10,000. The company had a turnover of £1 million last year and it is expected to be about the same in money terms this year. The company does not operate a bonus scheme. It has a jobbing department which contributes about £200,000 per year to the turnover. The profit last year was £37,100 and is expected to be £38,000 this year.

TYNEMOUTH CONSTRUCTION COMPANY LTD
Balance Sheet as at 31st May 19......

Authorised Capital 2,000 – £1 ordinary shares

2,000 – £1 ordinary shares	2,000	Plant	10,000
Loan by Rose	40,000	Freehold land	40,000
Secured loan	20,000	Goodwill	5,000
Revenue reserves	10,000	Debtors	40,000
Bank overdraft	10,000	Work in progress less payments	39,000
Sundry creditors	50,000	Stock	5,000
Future tax	4,000		
Current tax	3,000		
	£139,000		£139,000

Work in at present:

Jobbing	£140,000
40 revites	£320,000
3 private houses	£128,000
10 local authority houses	£350,000

The plant consists of:

3 15 cwt dumpers	Huts
4 5/3½ mixers	Scaffolds
1 10/7 mixer	1 7 tonne wagon
Joiners shop equipment – saw bench	2 15 cwt vans
mortice machine	1 car
overhead planer	All other plant is hired
thicknesser	
small machines	

The staff:

	Salaries
Managing Director John Rose	£14,000
3 General Foremen	£8,500 each
2 Travelling Supervisors	£8,500 each
1 Estimator/Buyer/Surveyor	£8,000
1 Office Manager	£7,000
2 clerks	average £4,000 each

SOLUTION

Would you recommend:
   a merger?
   a take over?
   a holding company?
   none of these?

If a new company is formed make recommendations about organisation, budgets, turnover, type of work, location of head office, plant policy, and any other points you may feel will be useful to the management.

Case 12
S. SANDS (NORTH EAST) LTD CASE

Tasval PLC are a large multi-national conglomerate. They operate a vigorous takeover policy and are interested in S. Sands (North East) Ltd.

You have been appointed to negotiate the takeover price and conditions of takeover. Tasval are prepared to pay cash, share for share or combination of both. The middle price of Tasval is 210p.

Your report should include the following:
(a) Take over conditions
(b) Proposed changes after the take over
(c) Proposed organisation including structure of the Board
(d) Proposed corporate plan for five years
(e) Proposed management by objectives. Tasval are keen to operate MBO in all its subsidiaries if possible.

Appendix A – Shows Profit and Loss Account to year ending 31st May
Appendix B – Shows Balance Sheet as at 31st May
Appendix C – Shows list of mechanical plant
Appendix D – Shows details of top management
Appendix E – Shows list of contracts

The Chairman and Managing Director of S. Sands (North east) Ltd is Walter Clayton. His grandfather started the business 80 years ago. Their Head Office and yard is in a suburb of Newcastle upon Tyne.

Walter Clayton is 58 years old, married with two sons, both of whom are in the business. Jim Clayton is 36 years old and is the Estimating/Surveying Director. John Clayton is 30 years old and is the Contracts Director.

Walter Clayton owns 70 per cent of the ordinary shares, his wife 10 per cent and each of the two sons 10 per cent each.

In accordance with the Articles of Association of the Company, the control of the Company must stay within the control of the family except by an extraordinary resolution passed in a company meeting.

The two sons are very close and generally support each other. Their mother will generally support Walter Clayton.

Walter Clayton and his wife are very keen that their two sons remain within the business.

There is adequate space in the Head Office for possible expansion if this is justified.

Turnover and Profit (loan) for the last 10 years is as under:

	Turnover	Net profit before tax
Last year – Year 1	£14,500,000	£400,000
2	£15,500,000	£250,000
3	£13,300,000	(£120,000)
4	£12,200,000	£21,000
5	£10,150,000	£301,000
6	£9,850,000	£202,000
7	£7,750,000	£775,000
8	£8,350,000	£412,000
9	£5,250,000	£421,000
10	£8,110,000	£810,000

Staff

The company employ approximately 400 direct operatives and 100 indirect staff consisting of clerks, storekeepers, typists, time-keepers, technical assistants, trainee site managers, general foremen and trades foremen.

The Company carries out its own joinery, brickwork, painting, plumbing, concreting and labouring.

All other work is sub-contracted.

The Company does not operate a bonus scheme.

Tradesmen are generally paid a spot bonus of £6 per day

Labourers are generally paid a spot bonus of £4 per day.

Planning

The Company does not do any pre-tender planning. Bar charts are used for long term planning. Short term planning is done in a very informal way and depends on the site manager in charge of the contract.

Estimating

Tender success rate last year was 1 in 6.

The Company does not use analytical estimating.

The Estimating Department receives no feed-back from sites.

The Company does not tender for civil engineering projects.

The geographical limitations are Durham, North Yorkshire, Northumberland, East Cumbria and Tyne & Wear.

S SANDS (NORTH EAST) LTD CASE 75

You have been allowed to interview anyone in the organisation and the following points have been noted.
1. Cost control is almost non-existent.
2. Industrial relations are good. Apart from the tea break strike in early 60's the Company have not experienced an industrial dispute.
   The Company is a member of the Building Employers Confederation
   Most of the operatives (approx 70 per cent) are in the UCATT Union
   Joint consultation does not take place
   There are four recognised shop stewards
   There are no safety representatives
   There is no safety committee.
3. Accidend Record – Two fatal accidents within the last two years. Both machine operators. Ten serious accidents and 20 non-serious (more than three days off work but less than two weeks), five occurred during the last two years.
4. The Company have a social club but only staff members attend any of the functions.
   Only staff members and their wives are invited to the annual dinner dance.
5. Work study is not practiced in the Company.
6. Private housing is not planned by any formal techniques.
7. There is no set buying procedure for housing.
   All material orders for contracts are placed at the beginning of the job and 'called off' by site managers.
   Most people agree that material waste is high.
   Taking off of materials is carried out by buying department.
8. The Company does not operate a planned maintenance scheme.
   Site Managers complain about the poor quality of service from the plant department.
9. The joiners shop has not made a profit for five years, but almost breaks even, e.g.

Last year Year 1	(£8,000) loss
2	(£6,000) loss
3	(£2,000) loss
4	(£150) loss

The Joiners shop is well equipped and is a good building and only supplies the Company. It does not do outside work except in very special situations.

You estimate that the shop is working to approximately 60 per cent capacity.

The joiners shop manager does all drawing required by the shop

and carries out the following:
Buying, pricing, cutting lists, supervision (he also has a joiners shop foreman) and general administration.
10. Marketing is not carried out by the Company.
The Company has a good reputation for quality.
Local Architects like to use the Company and always include the Company on tender lists.
11. The Quantity Surveying Director does not believe in pushing claims too much. He says it damages the relationship between the architects and the Company.
12. Site Managers are given a good deal of autonomy with little interference from Head Office.
Morale of the Site Managers is low. They feel they are treated as second class to Head Office staff and receive little support. They are highly critical of Head Office administration.
None of the Site Managers have received any formal training.
13. Walter Clayton is keen to sell the business. His two sons are not so keen unless guarantees of future employment are given or suitable compensation is paid. Mrs Clayton will accept any offer if it is considered fair by her husband. She would like to retire with her husband to their bungalow in the Canary Islands.
14. The Company does not use computers at all.
15. The Management blame the recession for the tight profit margins. They claim that other companies are doing the work at cost.

This will be the only construction company Tasval will have in their portfolio. They are anxious to expand if you consider it is possible. They will expect you to give some indication if this is possible and by how much.

It is the general policy of Tasval to put their own management into a Company when it is taken over, but this policy is flexible. They have other companies within the group associated with construction such as plant hire, brick making, builders' merchants, pressure jetting and instant accommodation. All of these companies are highly successful and produce a return on capital employed of between 38 per cent and 44 per cent.

Tasval PLC takes the total capital employed as share capital plus long loans plus any reserves. The holding Company adopts a policy of strict financial control, although Managing Directors are given a great deal of autonomy as long as they produce results.

Profit of Tasval PLC last year was £140 million. They are highly successful and have produced an increased profit each year for the last 20 years. The shares have a PER of 12 with a coverage of 3.2.

Appendix A

Profit and Loss Account for the year ending 31st May

	£		£
Office Salaries	686,000	Gross profit	1,995,700
Travelling Expenses	16,000		
Bank charges	15,000		
Car expenses	20,000		
Telephone	2,100		
Heating/Lighting/Cleaning	15,800		
Depreciation	80,000		
Repairs	18,000		
Rates	15,800		
Directors' remuneration	100,000		
Audit	17,000		
Advertising	30,000		
General expenses	180,000		
Interest on loan	400,000		
Profit before tax C/D	400,000		
	£1,995,700		£1,995,700

Appropriation Account

Tax	152,000	Profit brought forward	400,000
Dividend	80,000		
Reserves	168,000		
	£400,000		£400,000

Appendix B

Balance Sheet as at 31st May

Notes

Fixed Assets

			£
(1) Buildings and land			1,000,000
(2) Plant less depreciation			320,000
(3) Office furniture less depreciation			90,000

Current Assets

(4) Work in progress	6,000,000		
(5) Land for development	1,000,000		
VAT due	80,000		
(6) Debtors	3,020,000		
(7) Stock	300,000	10,400,000	

Less Current Liabilities

(8) Creditors	3,080,000		
Current tax	100,000		
(9) Bank overdraft	1,200,000	4,380,000	6,020,000
			£7,430,000

Financed By

(10) 2,000,000 £1 ordinary shares	2,000,000		
(11) 10% loan due for repayment 1984	4,000,000		
Revenue reserves	1,100,000		
(12) Capital reserves	178,000		
Future tax	152,000		7,430,000
(13) Turnover year ending 31st May — £14,500,000			

Notes to Balance Sheet

(1) Consists of Head Office and yard. Last valued 1976. Present value estimated at £1,400,000.
(2) Plant depreciated on reducing balance method of 25% per annum. 100% of plant written off in first year for tax purposes. Deferred tax now exhausted.
(3) As for note (2).
(4) (a) Generalal valued at cost. Profit of £100,000 has been assessed on the work in progress and included in gross profit.
   (b) It is estimated that £1 million of the work in progress will be paid within the next three months.
(5) Planning permission has been obtained. this land is valued at cost until a house has legal completion.
(6) It is estimated that £20,000 of this figure is bad debt, but has not been written off this year.
(7) Valued at cost on LIFO system.
(8) Includes £108,000 retention for sub-contracts which has not been released by client.
(9) Facility is £1,300,000.
(10) Authorised capital 2,000,000 £1 ordinary shares.
(11) From ICFC.
(12) From last revaluation of land and buildings.
(13) Calculated on sales plus closing work in progress minus opening work in progress.

Appendix C

Written down value of plant:

	£
1 no Hymac 580	15,000
2 no JCB 3	16,000
4 no 7 tonne tippers	40,000
10 no company cars	50,000
1 no 30 cwt hoist	800
2 no 5 cwt scaffold hoists	1,000
1 no 10 cwt mobile hoist	1,200
3 no 200 mixers	3,000
8 no 100 mixers	4,000
2 no 380 mixers	3,000
2 no porta saws	800
1 no bar bender	200
1 no reinforcement cutters	200
10 no 15 cwt dumpers	8,800
Loose tools	39,000
Formworks	30,000
6 no 30 cwt vans/trucks	36,000
1 no tower crane 30 m jib 35 m under hook independent	30,000
2 no 3 in. pumps	
1 no 2 in. pump	10,000
1 no 4 in. pump	
1 no low loader	10,000
Site Cabins	21,000
	£320,000

Note:
A physical check of the plant shows the above to be approximately 10% over-valued.

Appendix D

The Board consists of:

Walter Clayton — Chairman and Managing Director
Qualifications FCIOB
Past president of the Northern Region BEC
JP
Salary £32,000.

Jim Clayton — Estimating/Surveying director
FIQS
Salary £20,000.

John Clayton — Contracts Director
No qualification beyond 'O' levels (He got three)
Contracts Director for four years
Salary £20,000.

Jack Smith — Financial Director
No qualification
Commenced with the Company at age of 14 as a clerk.
He is now 53 years old. Knowledge of modern techniques is poor
Salary £14,000.

Jim Clark — Joinery Director
No qualification
Commenced as an apprentice joiner with W. Durham
Age 48
Salary £14,000.

Other executives consist of:
One buyer plus two assistants
Four Quantity Surveyors
Two Estimators
One Plant Manager
Three Contracts Managers
One Planning Engineer
One Office Manager

S SANDS (NORTH EAST) LTD CASE

Appendix E

Contracts on at present	Amount of Work planned to be completed next year
Notes	£
(1) Local Authority Housing	2,100,000
(2) Private Housing	600,000
Sports complex for Local Authority	1,600,000
Public House	900,000
Factory (1) for English Industrial Estates	1,800,000
Factory (2) for English Industrial Estates	1,900,000
Factory (3) for English Industrial Estates	2,100,000
(3) Factory (4) for Drum Industrial Estates	1,800,000
Revitalisation work for Local Authority	100,000
Refurbishment to local County Hall	2,100,000
Health Centre for county	1,200,000
Joinery works	600,000
(4) Small works	800,000
	£17,600,000

Notes
1. Expected to make a loss of £100K.
2. Houses sell between £30K and £40K.
   Plots cost approximately £10K.
3. Expected to make a loss of between £30K and £150K.
4. Jobs less than £60K.

Case 13
LAWNSWOOD HOUSING LTD

The Lawnswood Housing Company builds approximately 400 houses per year consisting mainly of four standard designs. The houses are built on quite large estates. The Company has also acquired some 100 hectares of land which they are developing an estate of individually designed houses on 1,000 square metre plots.

The company has a good reputation based on fair pricing, good quality workmanship, good design and completion on time. They also offer a good after sales service. The standard houses create no real problems but John Jones the Managing Director is not too happy about control of the special development. These houses are sold for between £100,000 and £300,000 each, often to influential people and he wants to make sure the high standards of the company are maintained.

The design for these houses is carried out by a very good imaginative architect who has his own practice but does all Lawnswoods work. Although the houses have a mark of individuality they all use the basic traditional trades.

The Problem
When a client has reserved a plot, which costs him a deposit of £1,000 he is put in touch with the architect. When designs have been agreed with the client, Lawnswood give the client a start date. The client requires not only a completion date, but stage dates as Lawnswood are paid by Building Societies for stage completions. the stages are:
1. Up to damp proof course
2. Roofed in
3. Plastered out
4. Completion.

Jones has commissioned you to devise a planning and control tool which will not only control production but also control cash flow. The line of balance technique is being used on their standard houses, but Jones cannot see it being suitable for the special development as houses are commenced as the clients come along and this is at irregular intervals.

Lawnswoods Planning Engineer could manage the estate work without any additional help in his section provided the technique used was reasonably uncomplicated. The technique should also be able to help the buying side, furnishing them with material requirement dates.

What would you propose?

Case 14
UK CONSTRUCTION LTD

UK Construction Ltd, a large building and civil engineering company, was taken over 18 months ago by a large multi-national company. the construction company operates in most parts of the British Isles with the exception of the Northern Counties (Tyne & Wear, Northumberland, Durham, Cleveland, Yorkshire and Cumbria).

UK Construction Ltd has decided to expand into the Northern Counties and has obtained a £2 million loan at 10 per cent payable as required by the holding company.

A separate company is to be set up called Northern Construction Ltd with an issued capital of £10,000 £1 ordinary shares to be issued at par.

Twenty-six hectares of land has been acquired with a separate loan of £1,000,000 from the holding company to build low price (£25,000–£30,000) housing units. The cost of the land did not include road charges. The loan is charged at 15 per cent per annum.

UK Construction Ltd has obtained 2 No contracts for office blocks to be built in Newcastle and it is intended to use the Northern Construction Ltd to build them. Contract No 1 is valued at £3 million and Contract No 2 is £4.5 million. Both contracts must commence within eight weeks time and be completed within two years.

You have been appointed Managing Director of Northern Construction Ltd. You have been given a large degree of autonomy provided you show results.

The following points should be noted:
(a) UK Construction Ltd has leased office accommodation for you at £7,000 per year. The accommodation has seven offices with room for a further 20 offices of the instant accommodation type. The lease is for seven years, the offices are situated on the outskirts of Newcastle.
(b) Planning permission has been obtained for 150 units with outline planning permission for a further 750 units. Roadworks have not yet commenced.
(c) The new company has no plant or material yard. There is no room for

these around the offices.
(d) Initial overdraft facilities have been set up for £100,000.
(e) Additional finance can be obtained from the holding company if a suitable case can be presented. They will charge 15 per cent interest.
(f) The land for development has been purchased by the holding company and will be transferred to Northern Construction Ltd as legal completion of houses has been made.
(g) No staff has been appointed.
(h) No plant, office furniture or vehicles apart from your own has been purchased.
(i) The Market mix of work is left to you.

Case 15
CASE – DISPUTE

J. Watson Ltd employs approximately 2,400 people throughout the country. They are at present involved in the construction of a city centre development. Work on the site has lagged behind and the costs are exceeding that which has been allowed.

A new project manager has been appointed and intends to tighten things up. He has issued a notice to the union as follows:
1. The Company will no longer pay for tea breaks. If a man takes a tea break the time will be deducted from his pay.
2. Anyone arriving late in the morning shall have his wages deducted as follows:
    8.00 – 8.15 lose 15 minutes pay
    8.15 – 8.30 lose 30 minutes pay
    8.30 – 8.45 lose 45 minutes pay
    If he arrives after 8.45 he will not be allowed on site.
3. Anyone coming on the site after lunch drunk will be dismissed.
4. If a man takes any time off during the week he will not be permitted to work on Sunday.
5. A 10 per cent retention will be made on the bonus for possible bad workmanship. This will be paid when the work has been passed by the Clerk of Works.
6. The company will carry out spot searches to reduce pilfering.
7. The bonus losses of one squad must be offset by the bonus of the squads in a similar trade.

The Union have sent a petition to management and would like the following points attended to immediately:
1. A bigger or extra canteen to be installed. the present canteen is too small. The present canteen holds 50 personnel. There could be 100 men on site.
2. The canteen to offer a better selection of food.
3. Safety boots to be supplied by the Company for all men on site.
4. The labour only plastering squads to be withdrawn.

DISPUTE CASE

5. The Company to pay £1/day attendance bonus.
6. The wages to be paid on Thursday morning rather than at 4.30 pm.
7. The site to remain open during the annual holiday for people who wish to work. (It is the intention to close the site for two weeks).
8. The holiday pay to be made up to the bonus level of the average of the four weeks prior to the holiday.

Note:
1. The average wage on the site (six days including Sunday) is £250/week.
2. The only 'labour only' people on site are plasterers.
3. The contract is seven weeks behind and is approximately half completed.

Negotiate a settlement.

Case 16
NORTHUMBRIAN CONSTRUCTION LTD

You have been called in as a management consultant to report on this company.

The Company was founded in 1930 as a small building company. Its head office is situated in a suburb of Newcastle upon Tyne. It operates within a 60 mile radius of its head office.

The company has expanded and now carries out general building, private house development and civil engineering. It is a private company. It has a reputation for good quality work. It employs approximately 1,000 people.

The turnover of the company is calculated on the basis of sales plus closing work in progress minus opening work in progress. Sales includes all in invoiced work and interim certificates. Work in progress is valued at cost.

The turnover for the last five years is as follows:

Year 1 (last year)	25 millions of pounds
Year 2	30 millions of pounds
Year 3	26 millions of pounds
Year 4	24 millions of pounds
Year 5	22 millions of pounds

The organisation of the company is shown in Appendix A.
The profit and loss account for last year is shown in Appendix B.
The balance sheet as at 31st May is shown in Appendix C.

NORTHUMBRIA CONSTRUCTION LTD CASE

The turnover last year is sub-divided as follows:

Housing	£6 million in 200 units	(£7 million)
Civil Engineering	£10 million	(£7 million)
General Building	£9 million	(£16 million)

Figures in brackets are previous years breakdown

The net profit before tax for the last five years is as under:

Year 1 (last year)	£250,000
Year 2	£600,000
Year 3	£600,000
Year 4	£200,000
Year 5	£400,000

Dividends declared:

Year 1	5p share
Year 2	10p share
Year 3	10p share
Year 4	5p share
Year 5	10p share

Plant is used for the companies use only and is not hired out. Hired plant is used to supplement owned plant.

The joiners shop produces only for the company. It is a good building adjacent to head office with adequate machinery. It has its own pressure treatment plant. The bulk of the stock figure in the balance sheet can be attributed to the joiners shop.

The age of the chief executives are shown in Appendix D.

The size of the civils contracts range from £200,000 to £2 million.

The size of the general building contracts range from £100,000 to £6.5 million. Anything under £100,000 is classified as small works.

The tender success rate at present is 1 in 6.

The company have one desk top computer which is used principally for wages, salaries and costs. Little else is done on it.

The accommodation at head office is adequate and could take an expansion of approximately 30 per cent if required.

The company has approximately £12 million worth of work on at present. Most of this work is scheduled for completion by next May.

Appendix A

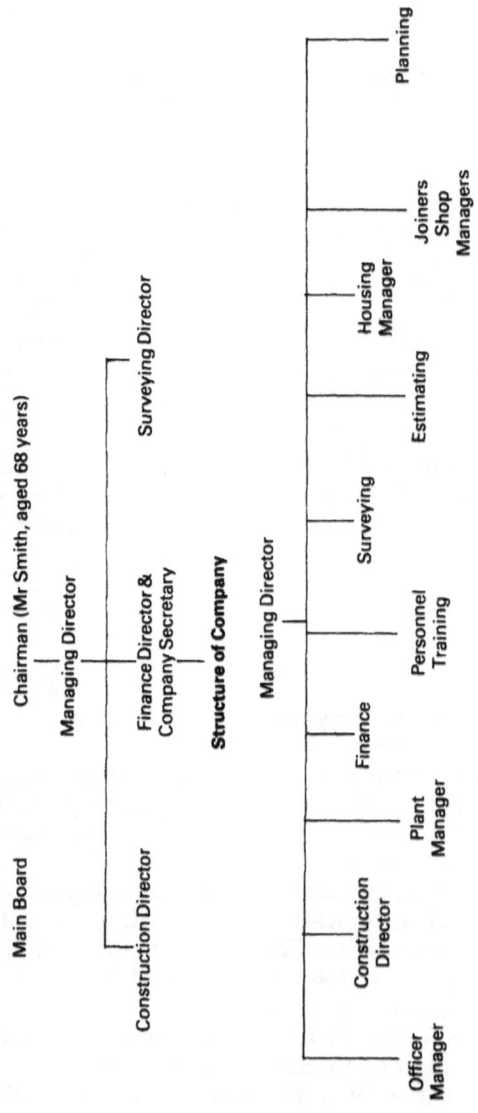

NORTHUMBRIA CONSTRUCTION LTD CASE

Sub-division of the Organisation

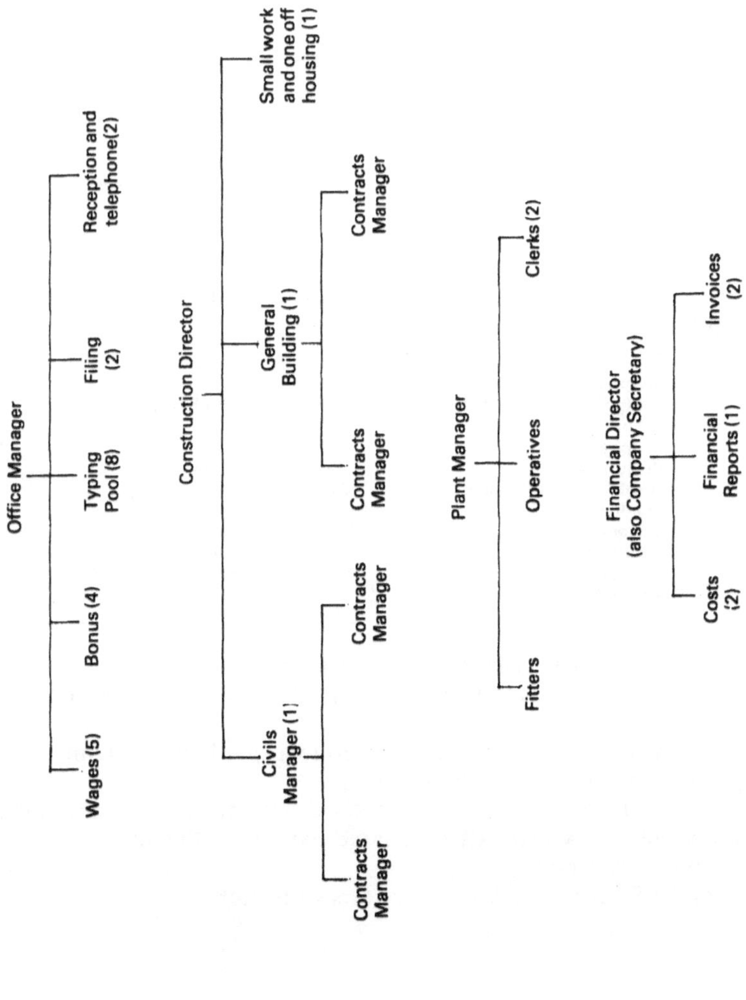

The breakdown of the £12 million is as follows:

Housing	£1 million
Civil Engineering	£7 million
General Building	£4 million*
*Includes small works and 'one off' housing.	

There has been a marked drop in housing sales. The average house price is £40,000.

The companay employs the following trades:

Joiners (bench and construction	Scaffolders
Bricklayers	Plumbers
Woodwork machinist	General Labourers
Concretors	Plant fitters and operatives
All remaining work is sub-let	

The following are some of the types of construction carried out over the last few years.

Shopping centre	Railway work
Sheltered Accommodation	Port of Tyne Authority work
Schools	Colleges
Hospitals	Offices
Health centres	Swimming pools
Local Authority Housing	Police station
Small road works	Public library
Sewers	College Library
Factories	Underpass to 6-lane road
Road bridges (2)	Sports complex
Metro stations	

It does not do re-furbishment work or jobbing.

Accounts for the joiners shop and plant department are not kept separate.

The bonus scheme has almost reverted to a spot bonus scheme.

The cost control system is ineffective.

Planning is done on the bar chart method with little short term planning being adopted. this depends to a large extent on the site manager in charge of that contract.

Appendix E gives details of interviews with staff.

Appendix B

Profit and Loss Account for the year ending 31st May

	£		£
Interest on loans	1,904,400	Gross profit	5,920,000
Office salaries	1,200,000		
Heating, lighting	25,000		
Office Expense	62,000		
Depreciation	240,000		
Director Remuneration	100,000		
Rates	30,000		
Insurance	25,000		
Auditors	21,000		
Car Expenses	180,000		
Travelling Expenses	60,000		
Bank charges and interest charges	1,820,000		
Repairs	2,600		
Net profit before tax	250,000		
	£5,920,000		£920,000

Appropriation Account

Dividend	100,000	Profit	250,000
Tax	130,000		
Reserves	20,000		
	£250,000		

Appendix C

Balance Sheet as at 31st May

	£000s	
Fixed Assets		
Land and Building Freehold (purchased 1948) – Revalued 1976	700	
Plant less depreciation	1,200	1,900
Current Assets		
Land for development	17,000	
Debtors	4,800	
VAT due	800	
Stock	1,900	
Work in progress less payments	7,200	
	31,700	
Less Current Liabilities		
Current tax	300	
Bank overdraft	51,000	
Creditors (including sub-contractor)	8,200	
	13,600	
		18,100
Financed By		
Shareholders funds		
1,000,000 £1 ordinary shares	1,000	
Revenue reserves	3,000	4,000
Deferred liabilities:		
12% loans	15,870	
Tax	130	
	16,000	
		20,000

Notes

Land for Development – Outline planning permission has been obtained for this. It is divided into 20 plots ranging from 6 unit plot to 100 unit plot.

Debtors – Excludes bad debts.

Loan – Is secured by a floating debenture and has seven years remaining of the 10 year term.

The largest shareholder is the Chairman, Mr Smith, who holds 40 per cent of the shares. The remaining 60 per cent are held by various members of the Smith family.

The overdraft facility £53 million. Land for development is at cost.

Appendix D

Age of executives	
Construction Director	56 years
Finance Director	58 years
Surveying Director	57 years
Office Manager	62 years
Plant Manager	63 years
Personnel Manager	64 years
Estimator (Chief)	56 years
Housing Manager	54 years
Joiners Shop Manager	58 years
Civils Manager	48 years
General Building	52 years
Small Works	56 years

Appendix E

The Chairman has stated that:-
a Profits are low.
b Labour turnover is high (300% last year).
c Communications are poor – the M.D. could not be very specific on this point except to say that too many things go wrong because people did not know what was expected of them and often misunderstood instruction.
d Job completions were often delayed.
e Morale in the company is low – this is shown by absenteeism, labour turnover, and general lack of co-operation.

The Chairman is pessimistic about the company weathering the recessional storm.

Interviews have been carried out by you and the main points which have emerged are:

Managing Director (age 60) Commenced his career as a joiner. He then progressed to foreman, general foreman, site manager, contracts manager and then to his present position. He has no formal qualification and apart from the odd days seminar has not received any 'off the job' training. He does not believe in 'off the job' training and says the only way to learn is by following experienced men 'on the job'.

The company does not use network planning as he considers it a waste of time. Bar charts are good enough he claims. When asked if the bar charts were successful he replied sometimes. He claimed to know about network planning as he had read articles on the subject but he had not received any formal training. He admitted his financial knowledge was weak but he says he has a good financial director and he can trust him with financial aspects. This also applied to the quantity surveying and the office work.

When asked how he spent his time he said the bulk of it was spent in visiting jobs, meetings, trouble shooting, meeting clients or potential clients and generally chasing people.

The M.D. has been with the company all his working life. He is a hard worker, he is always in the office (or site) by 8 am and rarely leaves before 6 pm. When asked about the low profit margins he blamed the recession and the keen prices other contractors were 'buying' the work for. Good site managers are hard to get he says. I have to keep them on their toes eg although the operatives finish at 3.30 pm on a Friday he expects the site managers to stay until 4.30. He will make some excuse to phone them up to ensure that they are there..

The following questions were asked:-
Q Are your people well motivated.
A I think so – they are paid enough and it is only money that will motivate people.
Q What is your salary.
A £16,000 per year plus £1,500 bonus plus car.
Q Would you still work for the organisation if your salary was dropped to £12,000.
A Yes.
Q Why.

NORTHUMBRIA CONSTRUCTION LTD CASE 97

A I enjoy work.
Q So you work for more than money.
A Yes, but I am the M.D. I would rather work than go on holiday.
Q How do you keep up to date.
A The job keeps me up to date.
Q What, in your opinion is the best way to manage people.
A Discipline – give them plenty of work and make sure they do it. If they don't they will be working for somebody else. Don't tell people too much – they might use it against you. I once lost a tribunal case because of that.
Q Have you had any claim for unfair dismissal against the company.
A At least 12 in the last 2½ years.
Q Who came off best.
A In 8 of the cases the employee. These tribunals don't know what they are talking about. It wasn't like this 20 years ago. You can hardly sack anyone now.
Q Do you hold regular meetings.
A No – they are generally a waste of time. I will soon find out if anything is wrong.
Q Do you use daily report sheets.
A No.
Q Have you thought about using management by objectives.
A I don't know anything about it.

Surveying Director
Q Is the company 'claims' conscious.
A Not very – we don't like to upset the architects.
Q Have you or your staff had any training in claims.
A No.
Q What is your relationship like with site managers.
A Generally not good. They are not cost conscious enough. One of them sacked a quantity surveyor who was working full time on his site. He said he had the authority. I said he hadn't. I tried to get him back, but failed. The site managers do not report the effects of variations enough. They look upon the Quantity Surveyor as an intruder.
Q Do you know what the work in progress ratio is at present.
A I don't know what you mean.
Q What is the average time to settle final accounts.
A Two years.
Q Do you employ micro computers at all.
A No. I don't really think they can help us.
Q How long have you worked for the organisation.
A 10 years.
Q Are you qualified.
A I passed the part II of the IQS examinations.
Q Do you think it would help if you were qualified.
A No.
Q What is your salary.
A £12,000 per annum. I also have a company car. I generally pick up a bonus of £800 per year.
Q How is the bonus calculated.

A I have no idea. It is worked out by the M.D.
Q Do you allocate surveyors to specific contracts.
A Yes, but I check all of their work.
Q Do you select your own staff.
A Yes, through the personnel department.
Q Are they all qualified.
A Some are – I am not sure how many.
Q Do you do any training in your department.
A I have one young trainee.
Q What control techniques do you use.
A We do a cost/volume reconciliation when we get a chance, but I am generally overworked.
Q Do you get involved in the selection of sub-contractors.
A Sometimes I do it, sometimes the buying department does it.
Q Who do you think should do it.
A Me, but the buyer thinks it should be him. I am not too bothered.

Site Manager
Q What is the level of morale amongst the site managers.
A Low.
Q Why.
A Difficult to say – I think the company is a little old fashioned. If you attempt to try any new ideas they are generally knocked back. I just give up trying.
Q Can you give examples.
A Yes, the planning is shocking – the techniques used are useless and the programmes are just used as a picture on the wall. We don't know the costs of a job until it is finished. Trying to get information on new construction techniques such as materials is almost impossible. The plant is badly maintained.
Q What would you say the reason for this is.
A Difficult to say – I think some of the top management are too long in the tooth. The status of the site manager is lower than a girl in the office. We are the people who get the job done but we only get kicked. We work longer hours than the men – we finish at 4.30 on Friday. I would generally finish at 4.30 on Friday anyway, but I resent being forced to do it. We often get paid less than the men because they are on bonus and we are not.
Q How long have you worked for the company.
A Two years. I worked 15 years for my last company until they wound up because of the M.D. retiring. I don't think I will stick this company out if I can find a better job.
Q Have you had any formal training.
A No. I wanted to attend the CIOB Site Management course but they wouldn't let me off work for 1½ hours per week.
Q Do you get a lot of support from head office.
A I get a lot of interference I don't get much support. I prefer to be given a free hand but there are times when I need help from the buying department, plant department and the contracts manager when I am having trouble with sub-contractors.

NORTHUMBRIA CONSTRUCTION LTD CASE

Q Would you say you are typical of the site managers.
A I would say the majority of them.
Q Have you the authority to hire and fire.
A Yes.
Q What could be done to help you to perform more effectively.
A Better training, more information, higher salary, clerical assistance on site, more support from head office, more sophistication for what I do and better techniques.
Q What would you do if you were the Chairman.
A Sack the board and all top management and start again.

Personnel Manager
Q What does the training officer do.
A Assists in the recruitment of apprentices Grant claims to the CITB.
Visits colleges where apprentices do their block release.
General clerical assistance in the department.
Q Have you carried out a training survey in the company.
A No the M.D. would not allow it.
Q Do you feel that the company has a commitment to the personnel department.
A No. I feel we are a necessary evil.
Q Does the company support further education.
A Only up to the age of 21.
Q How long have you worked for the company.
A Four years.
Q Where did you work before.
A In the printing industry.
Q How do the two industries compare.
A The printing industry is akin to jungle warfare.
Q Do you enjoy your work.
A I could if I got more support.
Q Does anyone have a job description.
A No – not that I know of.
Q What is your salary.
A £10,000 per year plus a car plus a bonus of about £300 per year.
Q Does the company do anything about management development.
A No.
Q Who decided on the present organisation structure.
A I don't know – I think it just grew.
Q What does the company do about management succession.
A We wait until someone departs and if there is no-one in the company to take their place we bring in new blood from outside.
Q Have you any idea what that will cost the company.
A A fair bit I imagine.

Construction Director
Age 56. Commenced his working life as a joiner than worked his way up through the ranks. He has been with the company most of his working life.
After interviewing him the following points emerged:-
1. He has little knowledge of costing, budgeting control and planning techniques.

2. He has received no formal training.
3. He complains about the quality of site managers.
4. He spends 65% of his time visiting sites.
5. He likes to be on site.
6. He has little time for quantity surveyors.
7. He, together with the personnel manager, interview all his staff.
8. He thinks the quality of work being produced is deteriorating.
9. The company should spend more money on plant.
10. He has little knowledge of material control.
11. He thinks the best way to manage people is by fear (if possible).
12. He is very boss centred.
13. He believes managers are born not made.
14. He considers the low profit is due to bad estimating, bad buying, no financial control and poor workmen.
15. He does not believe in financial bonus schemes as they produce poor quality work.
16. He has never heard of management development or MBO.
17. He believes all development is self development.

Financial Director
Age 58 years. Has been with the company since leaving school at 14. He studied at evening classes in accountancy but did not receive any major qualifications.

The following points emerged after interviewing him.
1. The company does not carry out any formal budgetary control.
2. The F.D. is unfamiliar with ratio analysis.
3. He blames the low profits on keen estimating and the keen market.
4. He does not do any training within his department.
5. He has a good knowledge of his duties as company secretary.
6. He is very methodical.
7. He believes in accuracy to the N'th degree.
8. He is thinking about introducing a better cost control system than they have at present.
9. He has not planned a successor to his job. He appeared rather perturbed when this was mentioned.
10. He complains that he does not get information in quick enough for financial returns. He did admit there was no set procedure.
11. He considered that the company must be efficient because it has been in existence so long.
12. He thought the possibility of it going out of business would be unthinkable.
13. He thought he was too old to introduce many new ideas.
14. He thought the desk top computer could be used more effectively.
15. He would not like to retire early. He would like to go on until he is 65 as he considers his work a very important part of his life.

Produce a report to the Chairman of the Company showing how the situation can be improved.

For Product Safety Concerns and Information please contact our EU representative GPSR@taylorandfrancis.com
Taylor & Francis Verlag GmbH, Kaufingerstraße 24, 80331 München, Germany

www.ingramcontent.com/pod-product-compliance
Lightning Source LLC
Chambersburg PA
CBHW021515120526
44766CB00007B/380